Democratizing Artificial Intelligence with UiPath

Expand automation in your organization to achieve operational efficiency and high performance

Fanny Ip

Jeremiah Crowley

BIRMINGHAM—MUMBAI

Democratizing Artificial Intelligence with UiPath

Publishing Product Manager: Ali Abidi

Senior Editor: David Sugarman

Content Development Editor: Joseph Sunil

Technical Editor: Rahul Limbachiya

Copy Editor: Safis Editing

Project Coordinator: Aparna Nair

Proofreader: Safis Editing

Indexer: Manju Arasan

Production Designer: Jyoti Chauhan

Marketing Coordinator: Priyanka Mhatre

First published: April 2022

Production reference: 1080422

Published by Packt Publishing Ltd.
Livery Place
35 Livery Street
Birmingham
B3 2PB, UK.

ISBN 978-1-80181-765-3

www.packt.com

Foreword

The mere mention of the term AI creates a cascade of reactions. We have all heard about aspects of the technology, some real, some prospective, and this has evoked a spectrum of views. Some believe the application of this capability is a giant leap forward to help solve some of society's most perplexing problems. Others are certain that it is opening the door to the marginalization of humanity and the subjugation of people to the whims of software. Regardless of your viewpoint, AI is rapidly becoming a widely used technology to help solve problems, both operationally challenging and predictively based, where large amounts of information are used to spot and specify trends and create valuable outputs.

There are those of us who have been involved in this area of technology for years, who have a growing number of those pesky gray hairs appearing on our heads and have seen several new capabilities appear and grow to become part of the new normal. From the perspective of time, these new capabilities tend to follow a clear, general trend. A new technology is developed and applied in a few cases with highly impactful results. As more people learn of this, they begin to imagine broader adoption with the promise of a quantum leap forward in solution design. Whether as a result of over-enthusiastic ambition or a lack of application expertise, there comes a bit of a resetting of expectations, a real-world reset if you will. However, through these initial phases of adoption, there is a growing number of dedicated professionals who are learning the skills necessary for both envisioning the use of the new technology as well as seeing its application in a broad and growing domain that expands the impact of this technology. This is typically when the rapid and steady growth of adoption takes flight.

This cycle of *Introduction – Over-hyped use or impact – Resetting of expectations – Democratization – Explosive growth* is a well-known trend. The democratization phase is one of the most impactful. When more practitioners understand and are skilled in the use of the technology, they can immediately begin to see possible applications in real-world situations. This phase accomplishes two things: one, it dramatically increases the skill base of practitioners, and two, it opens the aperture of deployment use cases and furthers the adoption of the new technology.

In this work, our authors have set out not only how to begin understanding and deploying this technology, but also to define and stratify the various levels of AI and clearly instruct the reader on how to use it. For the inquisitive reader, the road to expertise begins with the first steps and this book will guide you through those first steps. I encourage you to read this, understand and imagine the possible uses of AI, and let your own curiosity take you forward.

Foreword by Tom Torlone - Automation thought leader who has helped many companies scale their transformation programs.

Contributors

About the authors

Fanny Ip is a thought leader in automation, business transformation, and innovation. Currently, Fanny is the VP of the automation consulting innovation office at UiPath and leads a team of experienced management consultants in developing field-tested solutions for customers.

Before joining UiPath, Fanny co-led McKinsey's automation practice in North America and developed automation strategies for CXOs to deliver revenue growth and achieve a bottom-line impact. Before automation, Fanny has also led business transformation projects at PwC and Deloitte.

Fanny earned a BA in economics from the University of Chicago and an MBA from the Anderson School of Management at UCLA. Originally from Hong Kong, Fanny resides in Los Angeles with her family.

Jeremiah Crowley is a multi-disciplined developer with a wide technical and operational lens. His initial passion for software development and implementation led him to automation and process efficiency. He enjoys working on and solving problems, irrespective of their size and the technology involved.

Jeremiah received his BA in computer science from New York University. His work history includes the likes of EY and McKinsey & Company, where he assumed roles within their automation and service operations practices. He is currently a director within UiPath's automation consulting innovation office, assisting organizations in driving adoption and scaling automation by managing digital transformation initiatives and developing hyper-automation frameworks and methodologies.

About the reviewer

Azim Zicar is a low-code platform subject matter expert who helps businesses achieve digital transformation using RPA and Power Platform technologies. He is an experienced lead RPA developer certified by Blue Prism as an accredited professional developer and by UiPath as an advanced RPA developer. Azim is also a Microsoft Certified Power Platform Solution Architect Expert and has enabled multiple businesses to maximize and correctly utilize all of the different components of the platform. As a TOGAF certified enterprise architect, he can help you shape both business and technical strategies. He contributes to the community via public talks, workshops, and articles, all of which you can find on his website: Zicar Consultancy Ltd.

Table of Contents

3

Understanding the UiPath Platform in the Cognitive Automation Life Cycle

Section 2: The Development Life Cycle with AI Center and Document Understanding

7

Testing and Refining Development Efforts

Section 3: Building with UiPath Document Understanding, AI Center, and Druid

8

Use Case 1 – Receipt Processing with Document Understanding

11
AI Center Advanced Topics

Index

Other Books You May Enjoy

Preface

Artificial intelligence (**AI**) enables enterprises to optimize business processes that require cognitive abilities, are probabilistic, and have high variability and unstructured data in terms of documents, images, and voice. RPA developers and tech-savvy business users do not have a high barrier to using AI to solve everyday business problems. This practical guide to AI with UiPath will help AI engineers and RPA developers working with business process optimization to put their knowledge to work. The book takes a hands-on approach to implementing automation in business use cases in no time.

Complete with step-by-step explanations of essential concepts, practical examples, and self-assessment questions, this book begins by helping you understand the power of AI and gives you an overview of leveraging relevant out-of-the-box models. You'll learn about cognitive AI in the context of RPA, the basics of machine learning, and how to apply cognitive automation within the development life cycle. You'll then put your skills to the test by building three use cases with UiPath Document Understanding, UiPath AI Center, and Druid.

By the end of this AI book, you'll be able to build UiPath automation with the cognitive capabilities of Document Understanding and AI Center while understanding the development life cycle.

Who this book is for

AI engineers and RPA developers who want to upskill and deploy out-of-the-box models using UiPath's AI capabilities will find this guide useful. A basic understanding of robotic process automation and machine learning will be beneficial, but not mandatory, to get started with this UiPath book.

What this book covers

Chapter 1, Understanding Essential Artificial Intelligence Basics for RPA Developers, will cover the key **AI** concepts that are relevant in your daily work as an RPA developer.

Chapter 2, Bridging the Gap between RPA and Cognitive Automation, will explore in detail the benefits of adding cognitive automation to your **Robotic Process Automation** (**RPA**) toolkit.

Chapter 3, Understanding the UiPath Platform in the Cognitive Automation Life Cycle, will explore the UiPath platform to appreciate how using this platform can help you accelerate and amplify cognitive automation.

Chapter 4, Identifying Cognitive Opportunities, will focus on how to search for cognitive opportunities, and qualify automation opportunities, to ensure the opportunity is fit for cognitive automation.

Chapter 5, Designing Automation with End User Considerations, will focus on how to gather user goals and requirements, and learn how to incorporate these goals into the design of future state automation.

Chapter 6, Understanding Your Tools, will review the activities of Document Understanding, AI Center, and Computer Vision, while reviewing the Document Understanding framework and learning about the out-of-the-box models available with AI Center.

Chapter 7, Testing and Refining Development Efforts, will review how to prepare test data and test cases, and also how to train and increase model accuracy by closing the feedback loop with human validation and UiPath's built-in validation features.

Chapter 8, Use Case 1 – Receipt Processing with Document Understanding, will demonstrate how to build cognitive automation that can interpret the images of receipts using UiPath Document Understanding.

Chapter 9, Use Case 2 – Email Classification with AI Center, will demonstrate how to build cognitive automation that can classify the text of emails using UiPath AI Center.

Chapter 10, *Use Case 3 – Chatbots with Druid*, will demonstrate how to build a chatbot with Druid that can interact with UiPath automation.

Chapter 11, AI Center Advanced Topics, will cover advanced topics with UiPath AI Center, including Named Entity Recognition and deploying custom ML models.

To get the most out of this book

You will need to have experience building UiPath automation, ideally having completed UiPath's RPA Developer Foundation course. You should be familiar with working in UiPath Studio and UiPath Automation Cloud. You will need a version of UiPath Studio (2021.10+) installed on your computer.

Software/hardware covered in the book	Operating system requirements
UiPath Studio 2021.10	Windows 8.1+

You will also need an Enterprise License of UiPath Automation Cloud to continue with the book. A 60-day trial of UiPath's Enterprise Automation Cloud can be acquired at `https://cloud.uipath.com/portal_/register`.

If you are using the digital version of this book, we advise you to type the code yourself or access the code from the book's GitHub repository (a link is available in the next section). Doing so will help you avoid any potential errors related to the copying and pasting of code.

Download the example code files

You can download the example code files for this book from GitHub at `https://github.com/PacktPublishing/Democratizing-Artificial-Intelligence-with-UiPath`. If there's an update to the code, it will be updated in the GitHub repository.

We also have other code bundles from our rich catalog of books and videos available at `https://github.com/PacktPublishing/`. Check them out!

Code in Action

The Code in Action videos for this book can be viewed at `https://bit.ly/3DKdXul`.

Download the color images

We also provide a PDF file that has color images of the screenshots and diagrams used in this book. You can download it here: `https://static.packt-cdn.com/downloads/9781801817653_ColorImages.pdf`.

Conventions used

There are a number of text conventions used throughout this book.

Code in text: Indicates code words in the text, database table names, folder names, filenames, file extensions, pathnames, dummy URLs, user input, and Twitter handles. Here is an example: "Within the **variables** pane, set the `windowIdentifier` scope to the main sequence instead of `Do`."

Bold: Indicates a new term, an important word, or words that you see on screen. For instance, words in menus or dialog boxes appear in bold. Here is an example: "In **DefaultTenant**, click on the three dots and then click on **Tenant Settings**."

> **Tips or Important Notes**
> Appear like this.

Get in touch

Feedback from our readers is always welcome.

General feedback: If you have questions about any aspect of this book, email us at `customercare@packtpub.com` and mention the book title in the subject of your message.

Errata: Although we have taken every care to ensure the accuracy of our content, mistakes do happen. If you have found a mistake in this book, we would be grateful if you would report this to us. Please visit `www.packtpub.com/support/errata` and fill in the form.

Piracy: If you come across any illegal copies of our works in any form on the internet, we would be grateful if you would provide us with the location address or website name. Please contact us at `copyright@packt.com` with a link to the material.

If you are interested in becoming an author: If there is a topic that you have expertise in and you are interested in either writing or contributing to a book, please visit `authors.packtpub.com`.

Share Your Thoughts

Once you've read *Democratizing Artificial Intelligence with UiPath*, we'd love to hear your thoughts! Scan the QR code below to go straight to the Amazon review page for this book and share your feedback.

https://packt.link/r/1-801-81765-0

Your review is important to us and the tech community and will help us make sure we're delivering excellent quality content.

Section 1: The Basics

In this section, RPA developers will learn the relevant AI concepts, understand how cognitive automation works with RPA to increase the potential for automation, and gain an appreciation of the AI strategy and approach within the UiPath platform.

This section comprises the following chapters:

- *Chapter 1, Understanding Essential Artificial Intelligence Basics for RPA Developers*
- *Chapter 2, Bridging the Gap between RPA and Cognitive Automation*
- *Chapter 3, Understanding the UiPath Platform in the Cognitive Automation Life Cycle*

1
Understanding Essential Artificial Intelligence Basics for RPA Developers

In this chapter, we will cover some key **artificial intelligence (AI)** concepts that are relevant in your daily work as an RPA developer. We will discover where a **robotic process automation (RPA)** developer can make the most impact on implementing cognitive automation in RPA use cases without becoming a data scientist. We will also look at real business problems today that are solved by AI.

In this chapter, we will cover the following main topics:

- Understanding key AI concepts
- Understanding cognitive automation
- Exploring **out-of-the-box (OOTB) machine learning (ML)** models for RPA developers

By the end of the chapter, you will be equipped with common AI fundamentals, and you will be inspired by real-life examples to help you start thinking about how to apply AI to your potential use cases.

Understanding key AI concepts

You may have come across many terms when you started exploring the topic of AI. We will demystify AI and only present those concepts that are most relevant to you as an RPA developer. Please note that you may come across other material with slightly different definitions based on a different context.

Differentiating between artificial intelligence, machine learning, and deep learning

AI, ML, and **deep learning** (**DL**) are related but not the same. The following figure illustrates the hierarchy of these types of learning:

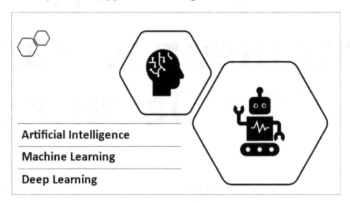

Figure 1.1 – AI, ML, and DL

- **AI**: This is equivalent to giving a machine or a robot the ability to think. It encompasses ML and DL.

- **ML**: This refers to how a machine or a robot learns to think through algorithms without explicit programming. ML is a subset of AI.

- **DL**: This refers to how an ML algorithm leverages artificial neural networks to mimic learning. DL is a subset of ML.

Next, we will look at three key considerations when choosing between ML and DL. They are listed here:

- Data requirement and availability
- Computational power
- Training time

The following figure shows a comparison of ML and DL:

	Machine Learning	**Deep Learning**
Data Requirement and Availability	• Can work with small datasets	• Must have large data sets
Computational Power	• Machines with a Central Process Unit (CPU) are sufficient	• Machines with a Graphics Processing Unit (GPU) are required
Training Time	• Short training time	• Long training time

Figure 1.2 – Comparison of ML and DL

In ML, the *features* of the studied subjects are fed into the algorithms for the machine to learn. We can think of *features* as us giving hints to the algorithm. This step allows for a smaller dataset, lower computational power, and less training time.

In DL, *features* are determined by artificial neural networks. It needs to work much harder to figure out the *features* and patterns to learn. As a result, it requires a large amount of data, high computational power, and a long training time.

Although DL is valuable, it is beyond the reach of most businesses to develop DL models to solve their business problems. Fortunately, many DL models have been pre-trained by companies with the time and budget to make them accessible to a large user base.

The implication of this option means that your role as an RPA developer is not to *create* these models. You, as the RPA developer, are the *trainer* of these models. It is important to understand the role of training in AI.

Appreciating the relevance of supervised learning, unsupervised learning, and reinforcement learning in AI

As we learned in the previous section, AI is about training a machine or a robot to think. Just like a human being, a robot needs to learn. There are three different types of learning for a robot.

The following figure gives some analogies for **supervised learning**, **unsupervised learning**, and **reinforced learning**:

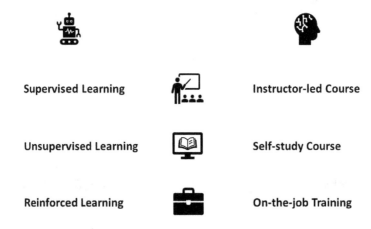

Figure 1.3 – Supervised learning, unsupervised learning, and reinforcement learning analogies

The following list explains the various analogies:

- **Supervised learning**: This is based on past data, and the *trainer* specifies the inputs to predict future outcomes. This type of training is analogous to an instructor-led training course. It requires the *trainer* to supervise the student or the model to achieve the desired learning outcome. Classification and regression are types of supervised learning methods:

  - **Classification** refers to the process of categorizing a given set of data into classes. For example, a set of pictures of different animals are fed into the ML model. Each picture is labeled with an animal name. The ML model is trained to identify animals from an image.

  - **Regression** helps in the prediction of a continuous variable. For example, a profit prediction ML model is an example of a regression model. Training data consisting of R and D, marketing, and administrative spending, geographic location, and profit is fed into the model. The ML model predicts the profit.

- **Unsupervised learning**: This relies on an algorithm to identify unknown patterns from data. This type of training is analogous to a self-study course. It requires the students or the model to synthesize the information to achieve the desired learning outcome. Clustering is a type of unsupervised learning method:

  - **Clustering** refers to the method used to find similarity and relationship patterns among training datasets, and then cluster those datasets into groups with similarities based on *features*. For example, the clustering technique is commonly used in market segmentation. The ML model looks at features such as sex, age, race, and geographic location to group customer groups into segments to better understand their buying habits.

- **Reinforced learning**: This uses a reward-and-punishment system to learn. There is no training data or *trainer*. The algorithm is improved over time based on feedback or reward and punishment. This type of training is analogous to on-the-job training. If the worker is doing the job well, the worker gains a pay raise or promotion. If the worker is performing poorly, the worker receives no raise or promotion. This is commonly used when no data or specific expertise is available.

Practical tips

AI platform providers have a mission to make AI accessible. Part of that mission is striving to develop product features to overcome the complex concepts of AI. Specifically, these are some notable democratization efforts in AI:

- Increased availability of pre-trained models to accelerate the time to result
- Simplification of the technical complexity of the ML training life cycle

We presented the key AI concepts in an easily digestible format. This overview prepares you to pick up an AI platform such as UiPath quickly. You will build, deploy, and maintain your first AI+RPA use cases in no time. You no longer need to spend years mastering AI to build a model from scratch. Instead, you are the *trainer* of the robots, teaching different skills that they need to master. Most importantly, you have tools that do the most complex tasks for you.

Now that you have a good understanding of the key AI concepts, let's explore **cognitive automation**, which is the combination of AI and RPA.

Understanding cognitive automation

Cognitive automation or **intelligent process automation** (**IPA**) refers to the use of AI and RPA together. It provides the machine or the robot with the brain (AI) and the limbs (RPA).

Although the general **software development life cycle** (**SDLC**) looks the same at a high level for RPA development and cognitive automation development, there are two important differences:

- The role of the RPA developer across the SDLC

- The final output of the RPA and cognitive automation life cycles

Let's now take a look at these differences in detail.

Understanding the expanded roles the RPA developer plays in the cognitive automation life cycle

An RPA developer plays expanded roles in the cognitive automation SDLC. A detailed comparison between a representative RPA SDLC and a representative cognitive automation SDLC is given in the following figure:

	Robotic Process Automation (RPA)	Cognitive Automation
Identify	• biz Analyst: End-to-end workflow requirements (input, process steps, and output)	• **RPA Dev: Data requirements for training / validation (availability, quality, and preparation) of ML models**
Build	• **RPA Dev: Develop RPA workflows**	• **RPA Dev: Train ML models with skillsets** • Data Scientist: Build ML models
Deploy	• **RPA Dev: Standard software deployment of RPA workflows**	• **RPA Dev: Easy plug-and-play deployment of out of the box ML models and custom ML models**
Consume	• Biz User: Each workflow relates to particular tasks and activities	• **RPA Dev: Easy Plug-and-play to drag ML skills into workflows**
Manage	• Administrator: Manage version control and governance of RPA workflows	• Administrator: Manage version controls and governance of ML packages
Improve	• **RPA Dev: Re-coding of bots per Biz User's requests**	• **RPA Dev: Continuous training of data** • Biz User: Validation stations to provide feedback to the models

Figure 1.4 – Differences in RPA developer roles in the RPA and cognitive automation SDLCs

In the RPA SDLC, an RPA developer is like a traditional developer for any other software package. In this, the typical sequence of the process is as follows:

1. The business analyst collects the end-to-end business requirements of a business workflow detailing inputs, process steps, and output.
2. The RPA developer codes the RPA workflow and tests the code.
3. The business user conducts a user-acceptance test of the RPA robot.
4. Finally, the RPA developer creates a package to deploy to the production environment.
5. Post-production, the administrator manages the operations of the RPA bots.
6. The RPA developer updates the code if the business user suggests enhancements or reports bugs.

The RPA developer plays a heavy role in selected steps of the RPA SDLC (build, deploy, and improve) by converting business requirements into RPA language.

In the cognitive automation SDLC, the RPA developer has a role in almost every step, which is described as follows:

1. The *RPA developer* collects data-specific requirements to prepare for ML model training/re-training.
2. The *RPA developer* does not usually build the ML model. Instead, the *RPA developer* either uses the ML model developed by the data scientist or uses an available OOTB model.
3. The *RPA developer* prepares the datasets for training and evaluation to train/re-train the ML model according to the specific use cases.
4. When the training result is acceptable, the *RPA developer* creates the ML package to deploy to the production environment.
5. The ML skills are then available for the *RPA developer* to plug and play in any RPA workflow.
6. Post-production, the *administrator* manages the operations of the RPA bots and the ML skills.
7. The *RPA developer* continues to re-train the model with new data points to improve the model.

In cognitive automation, an RPA developer plays a broader role across the SDLC as a *trainer* and a *data steward*.

Understanding the final output of the cognitive automation life cycle and the RPA life cycle

Another important distinction between RPA and cognitive automation is related to the characteristics of the final output produced. RPA configures RPA bots. Cognitive automation develops ML skills that are leveraged by the RPA bot. The following figure illustrates the differences in the expectations of an RPA bot and an ML skill in initial deployment to the stakeholders:

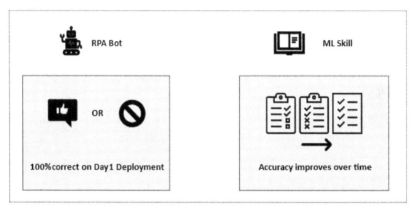

Figure 1.5 – Expectations of an RPA bot and an ML skill in the initial deployment

An RPA robot performs according to a set of rules set out by the RPA developer. The result is black and white. Only the correctly coded robot is deployed into production. The output of the cognitive automation life cycle is a trained ML skill combined with an RPA workflow. The ML skill is trained up to the acceptable threshold of confidence to be deployed into production. In almost all cases, the ML skill is not 100% correct when it is first deployed. The ML skill is expected to improve over time.

Practical tips

Businesses have seen the power and reap the benefits of automation through RPA. However, RPA has its limitations. RPA can only automate rule-based tasks, thus limiting the scope of a process it can automate. In addition, rule-based tasks are usually lower-value work. To move up the value chain, combining AI is essential for businesses to maintain a competitive advantage. Here are some of the key takeaways to bring to your leadership:

- Technology companies have simplified AI technologies to make them accessible for consumption. AI is no longer a tool that only data scientists can leverage.

- The existing RPA team can start incorporating AI without needing heavy investments in springing up a new team.

- There are impactful cognitive automation use cases throughout the organization.

- It is now time to give the machine or the robot a brain.

Now that you have a good understanding of cognitive automation, let's explore the most commonly used OOTB models that you can try as a beginner in AI.

Exploring relevant OOTB models for RPA developers

You have options when it comes to ML models. There are widely available OOTB models that you can use by re-training with your data. You can develop your ML models from scratch. Lastly, you can collaborate with the data scientists in your company on custom-built ML models.

In this book, we will provide tips on how you engage with these options. To begin, we recommend you start with the OOTB models. We will give you an overview of the most commonly used OOTB models in this section.

The commonly used OOTB models

OOTB ML models apply to a wide variety of use cases. They are pre-trained with a large amount of data. Some OOTB models can be retrained with your specific dataset, while others are not retrainable. Most automation platforms now include OOTB models. Selecting the right OOTB models can save you time and accelerate your project. The following figure illustrates the different categories of the OOTB models:

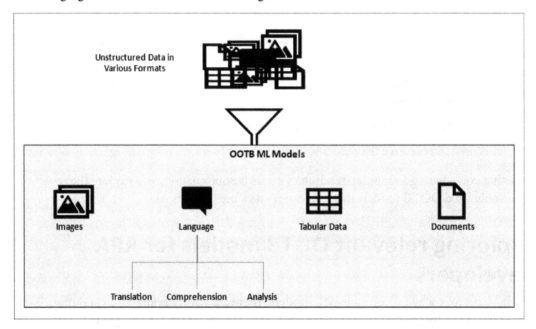

Figure 1.6 – OOTB ML models by category

These OOTB ML models convert various forms of unstructured data into a usable format. The usage of these models reduces reliance on humans to spend hours reading, processing, comprehending, and analyzing unstructured documents. Unstructured documents can come in the form of images, language, tabular text, and documents.

Let's take a closer look at each of these models:

- **Image analysis**: There are two image analysis OOTB models. The following figure summarizes the key characteristics of the two models:

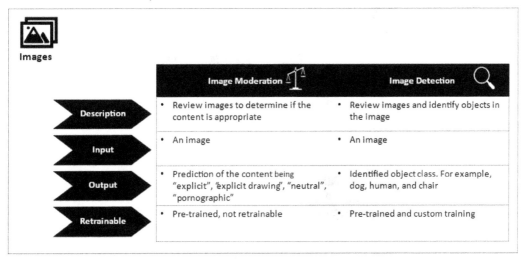

Figure 1.7 – Image analysis OOTB models

These two OOTB image analysis models are useful for many use cases that involve analyzing an image to determine the next steps. For example, the image moderation model is often used in social media feed moderation. The OOTB image moderation model reviews millions of images and flags images that may be problematic for humans to verify.

- **Language translation**: As the name suggests, language translation replaces the tedious work of translation from one language to another. The following figure summarizes the key characteristics of the model:

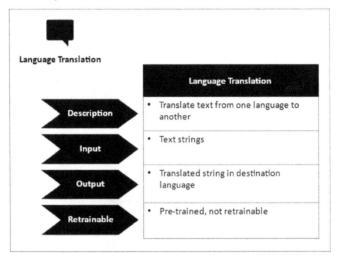

Figure 1.8 – Language translation OOTB models

This ML skill can be used in a variety of use cases and is commonly used in customer support. For example, many chatbots are powered by an OOTB language translation model to handle inquiries in different languages.

- **Language comprehension**: Language comprehension is complex. It refers to the ability to extract meaning from text, just like a human. The following figure summarizes the key characteristics of the three available models:

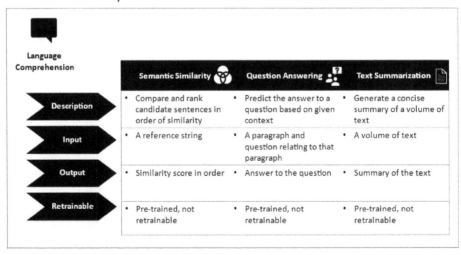

Figure 1.9 – Language comprehension OOTB models

Language comprehension ML models can mimic the thinking of a human and make inferences. They have widespread practical usage. For example, the semantic similarity OOTB model provides recommendations based on preferences indicated by the users. The question answering OOTB model is often used as a basis to build an automated **frequently asked questions (FAQ)** database. Finally, the text summarization OOTB model draws insights from books and articles.

- **Language analysis**: Language analysis refers to the skill of drawing meaning from text. It enables a machine or a robot to understand sentences and paragraphs. The following figure summarizes the key characteristics of the three kinds of models:

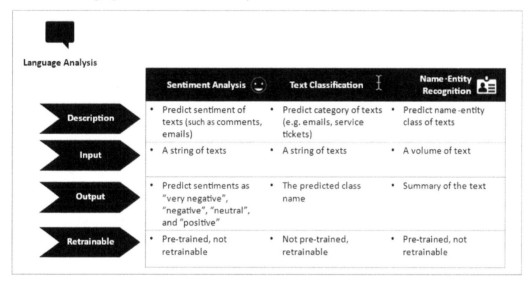

Figure 1.10 – Language analysis OOTB models

Language analysis ML models know how to draw context and relationships between individual words. They have widespread practical usage. For example, the sentiment analysis OOTB model is often used in managing emails from customers. The model prioritizes negative emails for humans to review. One popular usage of the text classification model is spam email classification. Finally, a named entity recognition model is often used to extract key parts from customer feedback.

- **Tabular data**: **Tree-based pipeline optimization tool** (**TPOT**) is a tool to find the best pipeline for your data. The following figure summarizes the key characteristics of the two available models:

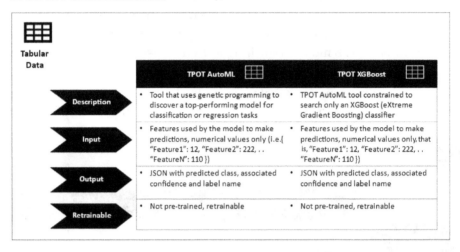

Figure 1.11 – Tabular data OOTB models

This OOTB tool automates the most tedious part of pipeline building. In addition, this is an introduction for a beginner to create a custom model.

- **Documents**: Processing documents is time-consuming and tedious. Many businesses spend many hours and a lot in human resources to digitize analog documents and extract structured information from them. The following figure summarizes the key characteristics of the three kinds of models:

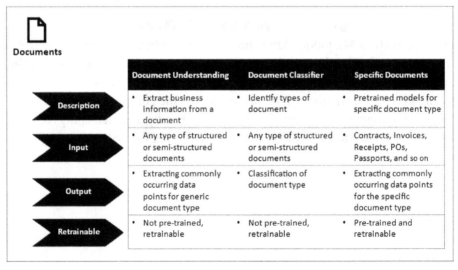

Figure 1.12 – Documents OOTB models

There are many documents on OOTB models available to tackle document digitization. They are often pre-trained with a large dataset of the relevant document type. They can be used to accelerate cognitive automation involving documents.

Practical tips

As we learned in this section, there are many OOTB models readily available. They have been widely used and proven to be effective. They are also easy to try. Think of a simple use case that involves AI skills and try your hand at any of the OOTB models mentioned in this section. Practice makes the theory you read in this book come alive.

Summary

In this chapter, you learned about the key AI concepts to start your immersion into AI. In addition, you learned about the power of cognitive automation to extend automation benefits and your role in cognitive automation implementation. Finally, you are now aware of the commonly used OOTB models for you to start hands-on exploration.

In the next chapter, we will dive into exploring the automation spectrum, the available technologies, and a framework to reimagine and solve a business problem with the relevant application of cognitive automation.

Further reading

- *MIT OpenCourseware – Artificial Intelligence*: `https://ocw.mit.edu/courses/electrical-engineering-and-computer-science/6-034-artificial-intelligence-fall-2010/index.htm`

- McKinsey's *An executive's guide to AI*: `https://www.mckinsey.com/~/media/McKinsey/Business%20Functions/McKinsey%20Analytics/Our%20Insights/An%20executives%20guide%20to%20AI/An-executives-guide-to-AI.ashx`

2
Bridging the Gap between RPA and Cognitive Automation

In this chapter, we will explore in detail the benefits of **cognitive automation** over pure **robotic process automation** (**RPA**). It is critical to deep dive into the **spectrum of office work** to understand the automation technology stack, the gap between RPA and cognitive automation, and the importance of human-machine coordination in making it work. We will also explore many real-life business problems for you to reimagine the art of possible automation potentials.

In this chapter, we will cover the following main topics:

- Understanding the spectrum of office work
- Exploring the gap between RPA and cognitive automation
- Designing human-machine collaboration with cognitive automation

By the end of the chapter, you will have learned how to approach business problems through a new lens of cognitive automation. You will know how to review automation opportunities holistically and how to design the optimal human-machine collaboration.

Understanding the spectrum of office work

We all wear many hats at work. If we look closely at our daily activities, each activity requires a specific skill set. Some are predictable and rule-based, while others require more cognitive power. To achieve the fullest extent of automation potentials, we must explore the details of the spectrum of work. We will focus on office work and not physical work in this book.

For any occupation, office work can be categorized into four categories.

For more information, you can refer to the following link:

```
https://www.mckinsey.com/~/media/mckinsey/business%20functions/
mckinsey%20digital/our%20insights/where%20machines%20could%20
replace%20humans%20and%20where%20they%20cant/where-machines-
could-replace-humans-and-where-they-cant-yet.pdf
```

The following diagram shows the breakdown of office work and its characteristics:

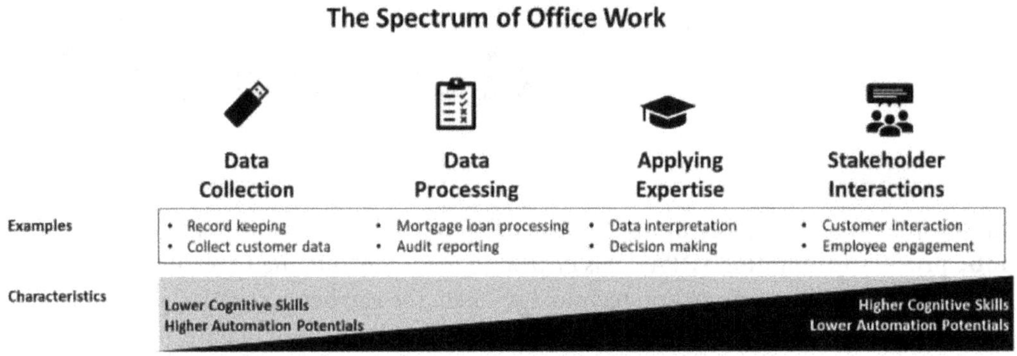

Figure 2.1 – The spectrum of office work

In the spectrum of office work, there are four categories of activities. They are **data collection**, **data processing**, **applying expertise**, and **stakeholder interaction**.

Data collection

Data collection is an activity that involves finding, extracting, and outputting data from many sources and systems to the destination system. In many cases, the most resource-intensive tasks are cleansing and formatting data according to a set of specific rules. These tasks generally have higher automation potentials. The following example illustrates different types of data collection efforts and challenges.

In the hire-to-retire process in **human resources** (**HR**), data collection takes up significant time. Employers spend a lot of time and resources in collecting employee life cycle data. Employers leverage **enterprise resource planning** (**ERP**) systems to minimize their efforts and time. However, there is still a lot of manual work and coordination that is not automated yet. Let's look at the employee onboarding process. The onboarding process includes the following data input from a variety of sources:

- A work authorization document is often a scanned copy of a physical document by the HR professional.

- Employee basic demographics data is usually collected via the employee portal.

- The background check result is transferred from the background check company to the employer electronically.

Data processing

Data processing is the manipulation of data elements into meaningful information. In data processing, rule-based straight-through processing is possible for clean data elements. Unfortunately, data is never 100% clean. There is always a need to process exceptions. Data processing is known as one of the first automated processes developed. With cognitive automation technology maturing, the automation potential to get rid of exception handling is increasing. The following example illustrates different types of data processing efforts and challenges.

Auditing is a high-paying profession. Both internal and external auditors have gone through extensive education and on-the-job training to hone their skills. However, they are also burdened with many data processing activities to complete audit procedures and produce audit reports. They spend limited time on problem solving and working with their clients to communicate and influence better business practices. Here are some resource-intensive data processing activities by auditors that have automation potential:

- Collect, follow up, and file requested documents for audits

- Perform procedural testing on systems according to controls requirements

- Download various system reports to ensure data integrity across systems

- Look for keywords and the associated context in documents for potential violations
- Review handwriting, approval stamps, and signatures in hardcopy documents

Applying expertise

Applying expertise is the ability to solve a problem with a collection of skills that are acquired through experience. There are two levels of expertise, as outlined here:

- **Routine expertise**: This relies on repeated procedures from the past to consistently solve problems
- **Adaptive expertise**: This is the ability to not just solve problems but solve them in new ways

There are many activities in our office work that require expertise. Expertise demands higher cognitive ability and is harder to automate. **Machine learning** (**ML**) is successful in automating routine expertise but is still developing its algorithms to handle adaptive expertise. The two following examples illustrate different types of applying expertise efforts and challenges:

- *Example 1*: Ingesting and interpreting invoices from paper or **Portable Document Format** (**PDF**) is a form of applying routine expertise. Before technology was available, companies had workers manually read invoices, determine the location and content of the invoice data, and then type it into the ERP system. For invoices, a data element such as an invoice number can be shown differently. A human worker understands that *Invoice Number*, *Inv. Number*, *Invoice No.*, and *Invoice #* are all referring to the data element invoice number based on many past transactions. Nowadays, ML and **natural language processing** (**NLP**) are available to apply this routine expertise to automate this activity.
- *Example 2*: For physicians, decision making can be routine if the patient's symptoms and the diagnosis are known. However, physicians encounter new or unexpected situations. **Artificial intelligence** (**AI**) is currently unable to determine what is relevant within new or unexpected situations. Adaptive expertise requires higher cognitive power and has lower automation potentials with available technology.

Stakeholder interactions

Stakeholder interactions refer to activities you have with your stakeholders. Stakeholders are individuals who have the power to affect and be affected by your actions. The goals of stakeholder interactions are to maintain the right touchpoint frequency, provide appropriate updates, address concerns, resolve conflicts, and engage interests. As you might expect, managing stakeholder interactions is complex. There are automation technologies to handle **question and answer (Q&A)**, service ticket prioritization, and automatic updates. However, there are activities in stakeholder interactions that involve empathy and emotions. These activities are still primarily performed by humans. The following example illustrates different types of stakeholder interaction efforts and challenges.

Proper customer engagement is vital to keep customers. Customer interactions come in many flavors: online shopping experiences, customer inquiries, customer complaints, product updates, and promotions. Chatbot technology is used to handle basic customer inquiries and reduce call volumes to live agents. Online shopping portals are designed with customers in mind to provide relevant information all in one place. Companies leverage ML and NLP technologies to triage customer emails and route the most urgent emails to the top of the queue, yet humans are still handling live interactions—be it for selling opportunities, conflict resolution, or white-glove service. It is not easy for machines to emulate emotions and empathy.

In this section, you have learned the details of the spectrum of office work. It is the first step toward exploring the fullest extent of automation potentials. Each activity requires a specific skill set. Some are predictable and rule-based, while others require more cognitive power. In the following section, we will explore the gaps between RPA and cognitive automation. In addition, we will look at the available technologies on the market. This exploration will help you determine what is available for you to fill the gaps.

Exploring the gap between RPA and cognitive automation

The spectrum of office work comprises activities that require different degrees of cognitive skills. Software companies have developed automation technology to tackle one or multiple areas.

The following diagram shows the available technologies across the spectrum of office work:

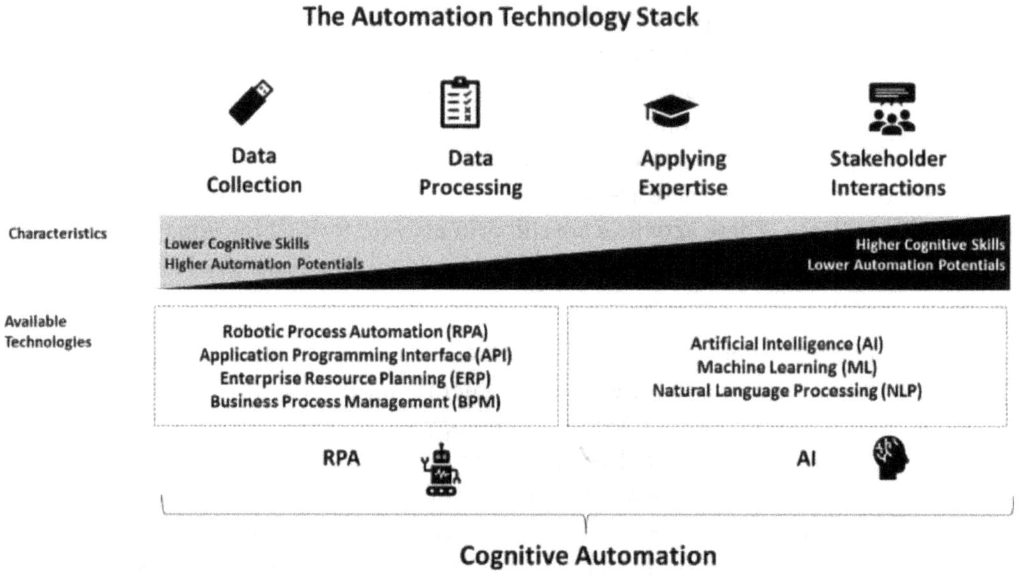

Figure 2.2 – The automation technology stack

RPA, **application programming interface (API)**, ERP, and **business process management (BPM)** belong to a class of automation technologies that reduce the manual efforts of rule-based digital activities. AI, ML, and NLP belong to another class of automation technologies. They reduce the manual effort of activities that require cognitive thinking.

As you learned in the preceding section, almost all activities have rule-based and cognitive components. Until recently, there were no platforms that allowed for seamless integration of these different technologies. Businesses look for a technology platform that allows users to take advantage of the different automation technologies. A new class of automation products called the **cognitive automation platform** has emerged. The cognitive automation platform simplifies the integration between systems and technologies. It also makes AI easy to develop and consume and bridges the gap between RPA and AI.

Practical tips

There are tremendous opportunities to apply appropriate cognitive automation to the spectrum of work to generate benefits. The most common mistakes and solutions of early cognitive automation adopters were these:

- They did not look at business problems holistically and were constrained by artificial boundaries such as **business units** (**BUs**). This hurdle required upper-level engagement to break through business silos.

- They wanted to test and deploy a specific technology and retrofit business problems to the technology. Many companies now have the mindset of business problems first and technology second.

- They were too optimistic about AI technology maturity and overestimated the AI capability. A better understanding of the different technologies and how they work would help set the appropriate expectation.

- They underestimated the importance of human elements, replacing delicate human touchpoints with technology without giving thought to the impact of such a replacement. Most recent digital transformation journeys include human-centric design in their discipline to avoid this problem.

By now, you should have a good grasp of the spectrum of work, the automation potentials, and the technology stack. Let's reimagine cognitive automation by perfecting human-machine collaboration.

Designing human-machine collaboration with cognitive automation

In the previous section, we explored the spectrum of office work and its relative automation potentials. Companies look for a more efficient and effective way to perform office work by leveraging automation. Office work does not comprise a single *activity* but a sequence of activities that make up a *process*. Full automation of the **end-to-end** (**E2E**) process may be the eventual goal. However, a more realistic goal is good *human-machine collaboration* to drive efficiency and effectiveness with currently available technology. Therefore, it is important to understand how humans and machines work differently and the best way to think through a human-machine collaboration when designing cognitive automation.

Demonstrating human-machine collaboration with examples

When we look at office work at a process level instead of at an activity level, we find that this is a collection of activities with different automation potentials. In most cases, humans are essential in completing the E2E process, and we must recognize the necessity of human-machine collaboration. The following diagram provides examples of human-machine collaboration in the E2E process:

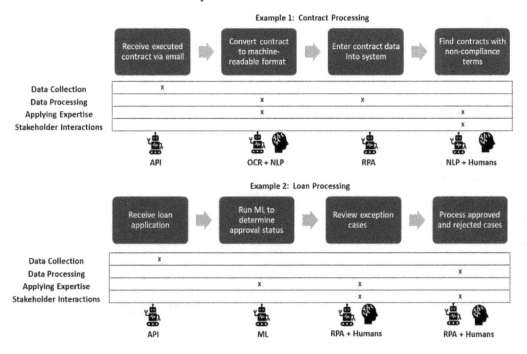

Figure 2.3 – Examples of human-machine collaboration

The first example of **contract processing** shows four steps that require different skill sets in the spectrum of office work, as follows:

1. *Receive executed contract via email (data collection)*: This step is automated. RPA is configured to route emails accordingly to the document understanding ML.

2. *Convert contract to machine-readable format (data processing and applying expertise)*: This step is partly automated. The document understanding ML (or **intelligent document processing**, or **IDP**) can understand most contracts. For contracts that have a low confidence level from the ML, humans validate and correct the interpretation. Machines and humans work together to complete this step.

3. *Enter data into the contract system (data processing)*: This step is automated. RPA completes the steps to log in to the system and input data elements extracted from the contract according to a set of rules.

4. *Find non-compliance contract terms and resolve (applying expertise and stakeholder interactions)*: This step is partly automated. The document understanding ML model has NLP to extract relevant contract terms from contract documents. However, humans are responsible for the interpretation and discussions with stakeholders to resolve non-compliance issues. Machines and humans work together to complete this step.

The second example of **loan processing** shows four steps that require different skill sets in the spectrum of work, as follows:

1. *Receive loan applications via a customer portal (data collection)*: This step is automated. An API routes data from the customer portal to the ML model.

2. *Run ML model to determine preliminary approval statuses (applying expertise)*: This step is automated. The ML model determines approval status based on past historical data. The ML is fine-tuned over time with new training data to ensure prediction performance. RPA routes preliminary approval statuses to either the exception case queue or the processing queue.

3. *Review exception cases (applying expertise and stakeholder interactions)*: This step is partly automated and is an example of a human in the loop with assisted RPA. Loan officers review cases that have lower confidence from the ML model. RPA pulls additional data for loan officers as reference material, along with the exception cases. Loan officers leverage their experience and the additional data to make final approval decisions. Machines and humans work together to complete this step.

4. *Process approved and rejected cases (data processing and stakeholder interactions)*: This step is partly automated. RPA processes the loan applications according to the approved and rejected workflows. In some cases, loan officers contact the customers to review the decision and make recommendations. Machines and humans work together to complete this step.

Applying differences between how humans and machines work in cognitive automation design

Humans and machines behave and perform differently. The following figure illustrates how humans and machines function in problem solving, memory, productivity, and emotions. These differences influence how we design cognitive automation to achieve a seamless human-machine collaboration:

Problem Solving	Heuristics: Depends on a rule of thumb, or a mental shortcut	Algorithm: Depends on a series of sets of defined steps
Memory	Short-term / Working Memory of ~7 pieces of information	Depends on RAM (Random Access Memory) capacity
Productivity	Duration of uninterrupted time	Any available runtime
Emotion	Ability to feel and empathize	No feelings and empathy

Figure 2.4 – Humans versus machines

Let's take a deeper dive into each of these areas, as follows:

- **Problem solving**

    Cognitive automation design implication: The first step is to assess the type of problem at hand to determine the best-suited solution. If the problems are predictable and have historical learnings, current automation technologies are more likely to be effective, whereas if the problems have many unexpected options, humans are more likely to be effective. With either humans or machines, the key to perfecting the algorithm is trial and error.

    In the following figure, you can see more information on how humans and machines solve problems:

| **Problem Solving** | ○ Humans solve problems using **heuristics**. Humans rely on the rule of thumb and the mental shortcuts to solve problems. This approach allows humans to solve problems efficiently without exhausting millions of possibilities. In addition, heuristics allow humans to come to an acceptable solution for unexpected situations. | ○ Machines solve problems using **algorithms** with output that is predictable and reproducible. Algorithms are a series of finite defined steps. Algorithms are best for predictable inputs and outputs. |

Figure 2.5 – Problem solving: humans versus machines

- **Memory**

Cognitive automation design implication: Humans process records one by one because of the limitation of short-term working memory. Machines have the choice of processing records one at a time or in batches. Using the same example, the design for machines is different than for humans.

- For humans, the pattern could go like this:

  - Row 1: Step 1, Step 2

  - Row 2: Step 1, Step 2

  - …

  - Row 100, Step 1, Step 2

- For machines, the pattern might look like this:

  - Step 1: Row 1, Row 2, …, Row 100

  - Step 2: Row 1, Row 2, …, Row 100

In the following figure, you can see more information on the differences between machines and humans in terms of memory:

Memory

○ Humans have **short-term working memory**. Short-term working memory holds approximately seven pieces of information. If there are a hundred rows of records to process and each row has five elements, humans can hold the information for only one to two rows at a time.

○ Machines memory depends on **RAM capacity**. Using the same example, machines can hold all rows of information for the next step of processing.

Figure 2.6 – Memory: humans versus machines

- **Productivity**

Cognitive automation design implication: Since humans require an uninterrupted duration of time, it is an important design principle to ensure freed-up capacities are in a contiguous block of time. In practice, this means moving from sequential steps in the pre-automated stage to a batch mode in the post-automated stage.

In the following figure, you can see more information on how productivity varies between humans and machines:

Productivity

○ Humans **require an uninterrupted duration of time** to be productive. Resuming focus on a task after an interruption takes time. A free-up capacity of one hour can be impactful depending on how this one hour is distributed. If sixty 1-minute increments make up the one hour, the free-up capacity is not productive. However, if one 60-minute increment makes up the one hour, the free-up capacity is productive.

○ Machines **work on any available runtime**. Using the same example, a free-up capacity of one hour of run-time is impactful regardless. Machines can fill up the run time with tasks for sixty 1-minute increments or one 60-minute increment.

Figure 2.7 – Productivity: humans versus machines

- **Emotions**

  Cognitive automation design implication: Technology has advanced to be able to mimic some feelings and empathy. However, the stakes are high if we apply automation in situations where maintaining a good stakeholder emotional state is important. During the cognitive automation design phase, we need to weigh the efficiency gain of replacing emotional responses with machines against the damage of having a wrong emotional response by machines.

  In the following figure, you can see more information on how humans and machines differ in their emotional responses:

Emotion	o Humans **have feelings and empathy.** Humans can react to emotional responses from stakeholders that are expected and unexpected.	o Machines **cannot feel and empathize on their own.** Language Analysis Machine Learning and Natural Language Processing can detect sentiment and understand the context to respond to expected scenarios.

Figure 2.8 – Productivity: humans versus machines

Now, let's go back to the example we discussed earlier and apply these design principles. The following diagram illustrates the future-state design using cognitive automation design principles:

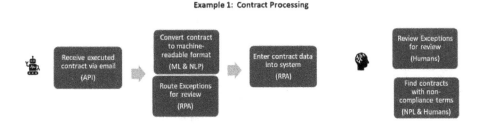

Figure 2.9 – Cognitive automation design illustration

In the contract processing example in the prior section, there are four sequential steps in the current state. In the future-state design, the process is regrouped into two. One is for the machines to complete, and one is for the humans to perform. We have applied each of the cognitive automation design principles to make the future-state design, as follows:

- **Problem solving**: This process stores contracts in the system and resolves non-compliance issues. Most activities are either rule-based or have a predictable outcome with sufficient training data. The current automation technology stack can automate most of the steps.

- **Memory and productivity**: Since humans and machines think and perform differently, the future-state design regroups activities into human activities and machine activities. The machine activities remain sequential and can be run at non-business hours to optimize the runtime schedule. Human activities are discrete and can be performed when it is the most convenient. Humans can decide to process a few activities at a time or all at once without dependencies on machines. The future state is designed with human-machine collaboration in mind.

- **Emotions**: In this process, the only step that involves delicate stakeholder interactions is the non-compliance resolution step. Therefore, in the future-state design, this step remains with humans. However, NLP is used as much as possible to identify non-compliance issues for humans to review. Humans handle all stakeholder interactions. This is another form of human-machine collaboration design.

Practical tips

Cognitive automation is a powerful tool to extend automation potentials. When human-machine collaboration is done right, it improves the **user experience** (**UX**). We have shown you the spectrum of office work and cognitive automation design principles.

Pick a past RPA project you have completed. Review the process to see if it is a cognitive automation candidate. If it is, apply the cognitive automation design principles you've learned about. Design a future state and compare it against what you have done with RPA alone. You will find that it is easy to extend automation benefits with just some tweaks on how you view the business problem and how you solve it with automation technologies.

Summary

In this chapter, we covered all aspects of cognitive automation. We understood the spectrum of office work and its characteristics. We explored the automation technology stack and the cognitive platform that fills the gap. We also learned the best practices for designing human-machine collaboration with cognitive automation.

What you learned in this chapter is critical for you to explore automation in a brand-new way to extend current automation benefit potentials. In the next chapter, we will conduct an overview of UiPath's E2E platform in the cognitive automation life cycle. We will explore UiPath Document Understanding, UiPath AI Center, and Druid chatbot.

3
Understanding the UiPath Platform in the Cognitive Automation Life Cycle

In this chapter, we will explore the UiPath platform. It is one of the leading **automation platforms** on the market. It is critical to understand the elements in the UiPath platform to appreciate how using this platform can help you accelerate and amplify with cognitive automation. We will also provide an overview of three key AI products in the platform that you will explore in detail in this book.

In this chapter, we will cover the following main topics:

- Understanding the critical success criteria in choosing a cognitive automation platform
- Introducing UiPath's end-to-end cognitive automation platform
- Getting to know UiPath Document Understanding
- Getting to know UiPath AI Center
- Getting to know the UiPath chatbot with Druid

By the end of the chapter, you will fully appreciate the end-to-end UiPath automation platform. You will understand the specific cognitive automation features each product has. You can reference the relevant chapters in this book to get hands-on experience and examples to use the respective product.

Understanding the critical success criteria in choosing a cognitive automation platform

A few years ago, automation and **Robotic Process Automation** (**RPA**) were synonymous. Many companies began their automation journey with RPA alone. In addition, IT ran RPA in the backroom. Over the years, automation technologies have evolved. Nowadays, companies use RPA, **Machine Learning** (**ML**), AI, chatbots, and NLP in their cognitive automation programs. Also, it becomes apparent that robots cannot do it alone. Collaboration between humans and robots is the key to unlocking the potential of automation.

It is important to select the right platform. You will learn the critical success criteria in choosing an automation platform that can support the **cognitive automation** transformation.

Guiding principles of a versatile automation platform

The following diagram shows the guiding principles and benefits of a powerful automation platform:

Guiding Principles of a Powerful Automation Platform

Guiding Principles	Details	Benefits
♲ Reusable	Out-of-the Box (OOTB) templates	Democratization
🚏 Simple	Low code, no code	Democratization
☞ Open	Simple integrations with legacy and best-of-breed applications	Flexibility
⚙ Powerful Configuration	With little to no customization required to tailor diverse needs	Speed to Market
🔒 Secure	Built-in governance for enterprise-grade protection	Enterprise Grade and Scale
↕↔ Sustain Change	Seamless upgrades	Low-Effort Maintenance

Figure 3.1 – Guiding principles of a powerful automation platform

The guiding principles are as follows:

- **Reusable**: The UiPath platform has many built-in **Out-of-the-Box (OOTB)** templates and a robust management feature of these reusable components to accommodate users with varying technical capabilities. UiPath Marketplace houses these reusable components from UiPath, partners, and community contributors.

- **Simple**: The UiPath platform always prioritizes usability. The platform is low-code/no-code to allow the democratization of the end-to-end life cycle, from developing and running to managing automation. It also allows for the democratization of various automation technologies, from RPA and document understanding to ML.

- **Open**: The UiPath platform allows for integration with other best-of-breed automation technologies as UiPath understands it may not be the only choice. As a platform provider, UiPath makes it easy for customers to put together the best-fit technology stack and operate seamlessly within the platform.

- **Powerful Configuration**: The UiPath platform is a low-code/no-code platform. It has powerful configuration capability to allow simple to complex functions to be configured effortlessly.

- **Secure**: The UiPath platform is designed for enterprise usage. Every feature set always ensures security and scalability. The UiPath platform conforms to the following industry standards: ISO/IEC 27001, Veracode continuous certification, and SOC 2.

- **Sustain to Change**: The UiPath platform automates processes. Many external changes cannot be controlled within the customers' four walls. The UiPath platform has features with built-in preventive controls to minimize the impact from unexpected changes. UiPath encourages the use of REFramework to ensure robust code design and maintainability. It also promotes the use of reusable components to create a separation of concerns. By incorporating object repository elements into discrete libraries for each application, this leaves only logic and no UI interaction within the main UiPath project. Lastly, UiPath Test Suite conducts regression testing to identify unexpected changes.

Design principles for human-machine collaboration in cognitive automation

As we discussed in *Chapter 2, Bridging the Gap between RPA and Cognitive Automation*, collaboration is key to extending the automation potentials to cognitive automation. The following diagram shows the key design principles that UiPath has followed to enable seamless human-machine collaboration on their platform:

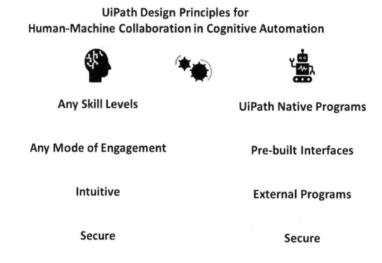

Figure 3.2 – UiPath design principles for human-machine collaboration in cognitive automation

The design principles in cognitive automation are as follows:

- **Humans**: The UiPath platform is designed for users with varying degrees of technical skill levels, from novices to professional developers. In addition, the UiPath platform allows users to engage with the platform in different modes – virtual machines, the web, mobile devices, and desktops. Most importantly, the interface between humans and machines must be intuitive and secure.

- **Machines**: The UiPath platform is an open platform that allows for UiPath's native programs, pre-built interfaces with the most popular systems, and external programs. UiPath has done an amazing job in the RPA world, but also chose to be the platform and infrastructure player for the AI portion. The open platform also comes with security and governance.

The UiPath platform follows the guiding principles of a versatile automation platform and the design principles of human-machine collaboration for cognitive automation. It is a platform that you can rely on to execute your transformational automation projects.

Introducing UiPath's end-to-end cognitive automation platform

UiPath was an RPA application provider. Over the last few years, UiPath has expanded its products and repositioned itself as one of the leading automation platform providers. As UiPath aims at a broader automation market and not just RPA, it makes a conscious effort to follow the design principles of human-machine collaboration in the platform for seamless cognitive automation.

Let's walk through the end-to-end product set of the UiPath platform. You will have a good understanding of how to leverage the product set in your transformation journey to maximize the potential of automation.

For more information, you can refer to the following URL:

```
https://www.uipath.com/product
```

The following diagram shows the UiPath end-to-end automation platform:

UiPath End-to-End Automation Platform

Cycle Stage	Discover	Build	Manage	Engage	Run
Goals	Discover everything worth automating and manage your org-wide rollout	Build automations quickly, from the simple to the advanced	Manage, deploy, and optimize automations across the enterprise	Engage people and robots as one team for seamless process collaboration	Run automations through reliable and flexible software robots that can take on a huge range of tedious tasks
Products	• Automation Hub • **Process Mining** • **Task Mining** • Task Capture	• StudioX • Studio • **Document Understanding** • Marketplace • Integrations	• Orchestrator • **AI Center** • Test Manager • Insights • Data Services	• Assistant • **Chatbots** • **Action Center** • Apps	• Unattended Robot • Attended Robot
Guiding Principles	♻ ⚑ ⚙ ✍	♻ ⚑ ⚙ ✍	🔒 ✛	♻ ⚑ ⚙ ✍	🔒 ✛

Bold – Products with AI

Source: https://www.uipath.com/assets/downloads/Platform-brochure

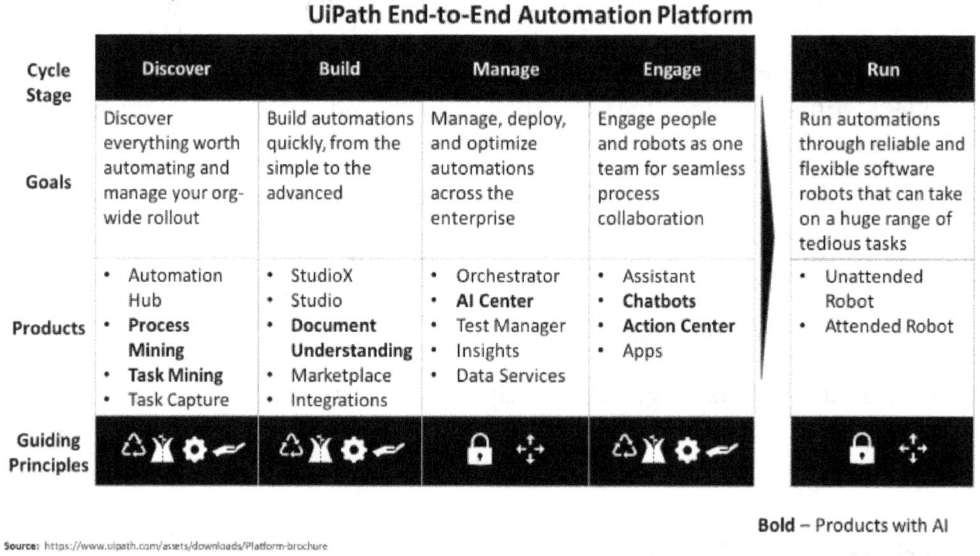

Figure 3.3 – UiPath end-to-end automation platform

The UiPath platform covers the full cycle of the automation life cycle: Discover, Build, Manage, Engage, and Run. Each pillar of the automation life cycle follows the guiding principles of a versatile platform:

- **Discover, Build, and Engage pillars**: UiPath products are designed to allow for a variety of skill levels to be able to use UiPath to start engaging, discovering, and building automation. The products always keep in mind pre-built templates, reusable components, low-code/no-code options, and flexible configuration.

- **Maintain and Run pillars**: UiPath products are also designed for enterprise scaling. The platform is secured and managed and has a governance hub to ensure things are secured from data to applications, can sustain constant changes internally and externally, can optimize resources on a large scale, and can monitor and proactively detect issues.

We will now check out each of the pillars in the following sections.

Discover pillar – discovering, evaluating, and managing automation use case pipelines

The **Discover** pillar of the automation development life cycle refers to the activities to find qualified automation opportunities. Traditionally, this phase is labor-intensive and subjective. The UiPath platform has four products for this pillar to make it efficient and objective. The following diagram shows how UiPath products target the different activities within the Discover pillar:

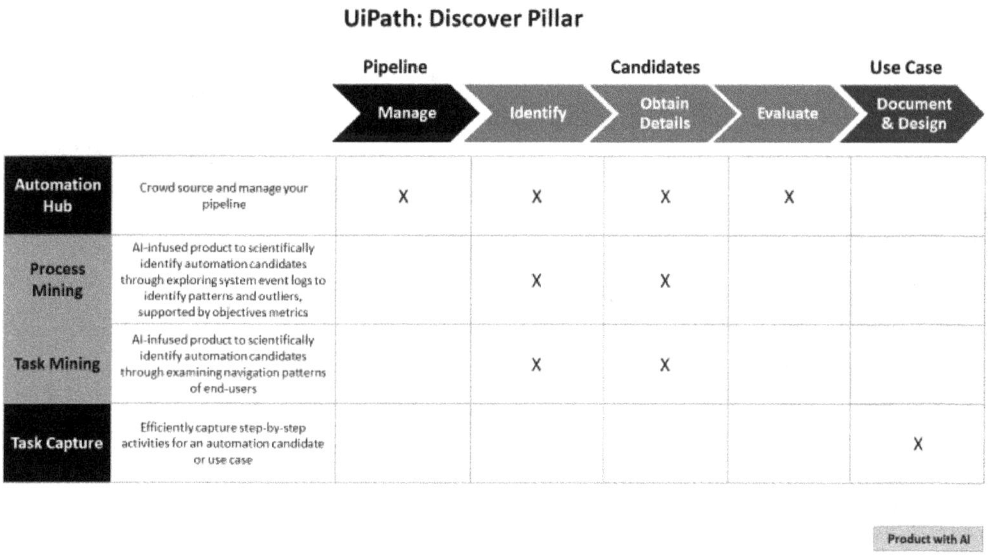

UiPath: Discover Pillar

		Pipeline	Candidates			Use Case
		Manage	Identify	Obtain Details	Evaluate	Document & Design
Automation Hub	Crowd source and manage your pipeline	X	X	X	X	
Process Mining	AI-infused product to scientifically identify automation candidates through exploring system event logs to identify patterns and outliers, supported by objectives metrics		X	X		
Task Mining	AI-infused product to scientifically identify automation candidates through examining navigation patterns of end-users		X	X		
Task Capture	Efficiently capture step-by-step activities for an automation candidate or use case					X

Product with AI

Figure 3.4 – UiPath: Discover pillar

The four products are **Automation Hub**, **Process Mining**, **Task Mining**, and **Task Capture**:

- **Automation Hub** is the keystone product in the Discover pillar. It is a tool for collaborative process identification, automation pipeline management, and a process repository tool.

- **Process Mining** and **Task Mining** are AI-infused products. **Process Mining** transforms data from your IT systems into visual interactive dashboards, understands root causes and risks, and identifies automation focus areas. **Task Mining** collects employee desktop data, comprising the screenshot and logs data upon each user action, to suggest a list of processes with high automation potential.

- **Task Capture** helps you dive deep into automation ideas, enabling you to quickly capture, enhance, and accelerate automation by sharing the specifics of your work.

Build pillar – developing automations

The Build pillar of the automation development life cycle refers to the activities for building automations quickly, from simple activities to advanced ones. Traditionally, the only way any development work can be done is through professional developers and to build from scratch. The UiPath platform provides a variety of ways to allow different skill levels to be able to develop their automations. The following diagram shows how UiPath products target the different types of developers within the Build pillar:

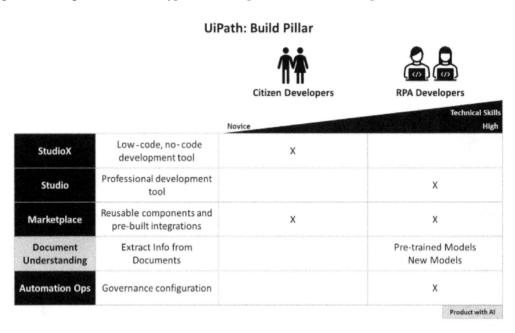

Figure 3.5 – UiPath: Build pillar

The five products are **StudioX**, **Studio**, **Marketplace**, **Document Understanding**, and **Automation Ops**. These products are designed specifically for citizen developers and RPA developers:

- **StudioX** is a no-code platform for non-technical citizen developers to automate their tasks. In addition, StudioX is part of the UiPath platform, which means the enterprise-grade governance and security infrastructure is in place to allow enterprises to have peace of mind.

- **Studio** is an advanced automation software that gives professional RPA developers the right automation canvas to build robots. Studio offers a comprehensive set of tools for designing complex attended, unattended, and testing automations.

- **Marketplace** is a library of pre-built RPA content. Both citizen developers and professional RPA developers can take advantage of the marketplace and integrations. The content includes native UiPath integrations and submissions by UiPath partners and individuals. All the content published on the marketplace undergoes an extensive certification process, making automation even more beneficial as the development time decreases without sacrificing the overall content quality and security.

- **Document Understanding** helps robots understand documents with AI. This product allows professional RPA developers to use pre-trained models to accelerate their use cases involving documents.

- **Automation Ops** is the central management console to keep automation safe. It allows companies to customize organization policies and tailor rules for groups and individuals.

Manage pillar – managing, deploying, and optimizing automations

The Manage pillar of the automation development life cycle refers to the activities to manage, deploy, and optimize automations across the enterprise. As an automation platform, UiPath products are tailored to manage all aspects, from bots, AI skills, and data to changes. In addition, it has a robust reporting engine to measure the effectiveness of the automation programs. The following diagram shows how UiPath products cover all aspects of managing the overall program:

UiPath: Manage Phase

		Bots	AI Skills	Data	Change Controls	Metrics & Reporting
Orchestrator	Command center to monitor and manage all aspects of bots	X				
AI Center	ML Ops for AI Skills		X			
Data Services	Manage data services across automation needs			X		
Test Suite	Proactively test expected and unexpected changes				X	
Insights	Metrics and reporting					X

Figure 3.6 – UiPath: Manage phase

The five products are as follows:

- **Orchestrator** is an enterprise-grade monitoring and administrative tool. UiPath Orchestrator allows the central orchestration of provisioning, deploying, triggering, monitoring, measuring, and tracking the work of attended and unattended robots.

- **AI Center** is an application that allows you to deploy, manage, and continuously improve ML models and consume them within RPA workflows in Studio. There are pre-built models, AI solution templates, multiple deployment options, and a drag-and-drop interface designed for RPA developers.

- **Data Services** enables RPA developers to easily access the data, variables, and tools they need across the entire UiPath platform easily. It integrates easily with any enterprise or legacy system, database, or custom app with a drag-and-drop storage interface.

- **Test Suite** is a test management tool for enterprise testing. It integrates into your ecosystem and adapts to your way of working, whether you use ServiceNow, SAP Solution Manager, Jira, and/or Azure DevOps. Test Suite has tight integrations with test management and ALM tools.

- **Insights** is an RPA analytics solution that tracks, measures, and manages the performance of the automation program. It has a library of curated dashboard templates, including robots, processes, queues, and business ROI. It also calculates the business impact of your automations and shares results quickly with key stakeholders across your company with self-service reports.

Engage pillar – facilitating human-robot collaboration in automations

The Engage pillar of the automation development life cycle refers to the activities to engage humans and robots as one team to collaborate. There are four products in the Engage pillar and they are particularly important for the cognitive automation we are working on – the human-machine collaboration. The following diagram shows how UiPath Engage pillar products engage humans and robots:

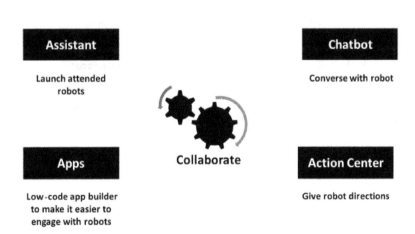

Figure 3.7 – UiPath: Engage phase

The four products are as follows:

- **Assistant** is a desktop launchpad for automation. It allows individuals to trigger an automation to assist with their day-to-day tasks. Assistant is managed, distributed, and governed centrally.

- **Chatbot** can intelligently trigger robots to do things. It automates conversations between humans and robots. Chatbots are often used for Q&A, notifications, form submission, reporting, or task tracking.

- **Apps** is a low-code application development platform that connects to data in any underlying system. It enables business users to create custom applications to facilitate human-robot collaboration.

- **Action Center** makes it easy and efficient to hand off processes between robots and humans. Action Center is used when automation includes decisions that a human should make, such as approvals, escalations, and exceptions.

Run pillar – running automations

The Run pillar of the automation development life cycle refers to the activities to run the attended and unattended robots. The following diagram shows how the UiPath products allow different modes of running the robots:

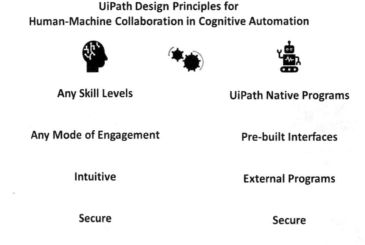

Figure 3.8 – UiPath: Run phase

The two products are as follows:

- **Unattended robots**: Unattended robots reside in virtual machines and run in the background. Unattended robots are used for rule-based, long-standing workflow automation use cases. They interact with humans via Action Center. In addition, AI skills can be added to unattended robots to perform cognitive tasks.

- **Attended Robots**: Attended robots work side by side with humans on their desktops and act as personal assistants to help with daily tasks. Humans access, schedule, and run attended automation in UiPath Assistant. The UiPath platform allows AI skills to be added to the unattended robots to perform cognitive tasks.

As we have seen in this section, UiPath is a powerful automation platform. You need to appreciate the full suite of products and what they are designed for as you think through how to leverage the full product suite for your automation needs.

Getting to know UiPath Document Understanding

UiPath Document Understanding is a powerful product that leverages AI to enable an RPA robot to understand documents. The product is flexible to accommodate extracting information from multiple document types. It allows you to create one single workflow that will extract data from a variety of documents.

The benefits of UiPath Document Understanding

The following diagram shows the benefits of using UiPath Document Understanding in the UiPath platform:

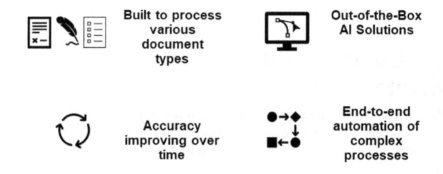

Figure 3.9 – UiPath Document Understanding: Benefits

UiPath Document Understanding provides four key benefits:

- *Built to process various document types and OOTB AI solutions*:

  - **UiPath Document Understanding** can handle a wide range of documents and a rich set of pre-trained ML models. The following diagram shows UiPath Document Understanding document types and pre-trained ML packages:

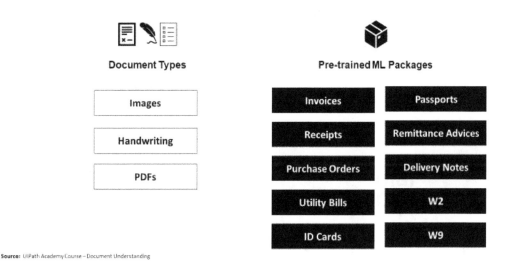

Figure 3.10 – UiPath Document Understanding: Document types and pre-trained ML packages

- **UiPath Document Understanding** can process a wider range of documents by leveraging its own understanding engine or the engines from its partner ecosystem. UiPath Document Understanding also has one of the richest sets of pre-trained ML packages on the market. The pre-trained packages include invoices (General, Australia, Japan, India, and China), receipts, purchase orders, utility bills, passports, ID cards, W2, W9, remittance advice, and delivery notes. UiPath also regularly adds new pre-trained packages.

- *Accuracy improving over time*:

  - **UiPath Document Understanding** also incorporates human validation effort seamlessly with UiPath Action Center and/or the UiPath attended robot. In addition, the corrected data is fed back into the ML to improve accuracy over time. The following diagram shows how UiPath Document Understanding validation and training works:

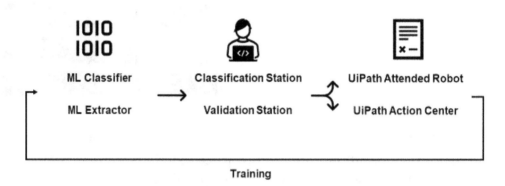

Figure 3.11 – UiPath Document Understanding: Validation and training

  - **UiPath Document Understanding** enables human-bot collaboration in optimizing the result. The models can be retrained automatically based on human input. The more you work with the model, the more effective it becomes.

- *End-to-end automation of complex processes*:

  - **UiPath Document Understanding** is not a standalone product. Combined with the rest of the UiPath platform, it can automate an end-to-end business workflow, as illustrated in the following diagram:

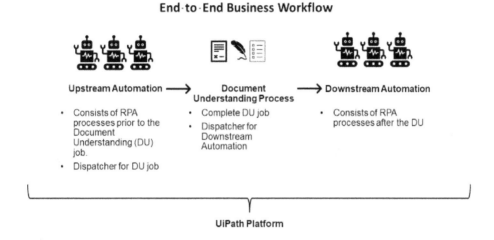

UiPath Document Understanding
End-to-End Business Workflow

Upstream Automation ⟶   Document ⟶ Downstream Automation
Understanding Process

- Consists of RPA processes prior to the Document Understanding (DU) job.
- Dispatcher for DU job

- Complete DU job
- Dispatcher for Downstream Automation

- Consists of RPA processes after the DU

UiPath Platform

Figure 3.12 – UiPath Document Understanding: End-to-end business workflow

- The end-to-end business workflow allows for maximum efficiency to manage overall workflow and robot utilization. It also prevents external issues from impacting Document Understanding jobs.

UiPath Document Understanding technical framework

Let's walk through the technical framework of UiPath Document Understanding. You will have a good overview of each component in the framework and how they relate to each other. You will be able to leverage UiPath Document Understanding in automating use cases with a document understanding activity. The following diagram shows the **UiPath Document Understanding framework**:

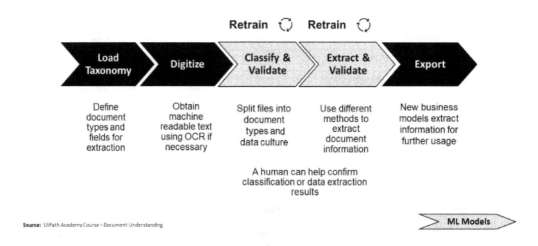

Figure 3.13 – UiPath Document Understanding

The UiPath Document Understanding framework includes five components. They are as follows:

1. **Load Taxonomy**: The first step is to define document types and fields for extraction to create a dedicated taxonomy structure in Taxonomy Manager.

   • **Use Case**: Extract information from travel and entertainment receipts for expense reporting.

     ◆ **Document type**: Receipts

     ◆ **Fields to be extracted**: Transaction date, Vendor, Total amount, Tax amount

2. **Digitize**: The second step is to turn a file into machine-readable content. This step requires the selection of an **Optical Character Recognition (OCR)** engine to digitize the content. UiPath Document Understanding has built-in integration with the most popular OCR engines. The OCR engines are UiPath Document OCR, OmniPage OCR, Google Cloud Vision OCR, Microsoft Azure Computer Vision OCR, Microsoft OCR, Tesseract OCR, and ABBYY Document OCR.

3. **Classify and Validate**: The third step identifies what types of documents the robot is processing. This step leverages AI/ML. There are three sub-steps to this step:

   - **Document classification**: This sub-step automatically determines the relevant document types. There are four OOTB classifiers: Keyword-Based Classifier, Intelligent Keyword Classifier, FlexiCapture Classifier, and Machine Learning Classifier. You can also create your own classifier to implement a custom algorithm.

   - **Document classification validation**: This sub-step allows for human validation using the **Classification Station**. There are two options when it comes to using the Classification Station: as an attended activity or as an Action Center task. This sub-step is optional but highly recommended if you are dealing with multiple document types in a single file or if you need 100% accuracy.

   - **Document classification training**: This sub-step passes the human-validated information back to the classifiers to improve their future predictions. This sub-step closes the feedback loop for classifiers that are capable of learning from human feedback. The available classifier trainers are Keyword-Based Classifier Trainer, Intelligent Keyword Classifier Trainer, and Machine Learning Classifier Trainer.

4. **Extract and Validate**: This step identifies specific information for extraction from the documents. This step leverages AI/ML. There are three sub-steps to this step:

   - **Data extraction**: This sub-step automatically extracts the relevant information from each document type as defined in the taxonomy. This step ensures that the configured extractors are called in the correct order, for the relevant list of fields, and for the specified page range of the file being processed. There are five out-of-the-box extractors: Regex-Based Extractor, Form Extractor, Intelligent Form Extractor, FlexiCapture Extractor, and Machine Learning Extractor. You can also create your own extractor to implement a custom algorithm.

- **Data extraction validation**: This sub-step allows for human validation using the **Validation Station**. There are two options when it comes to using the classification station: as an attended activity or as an Action Center task. This sub-step is optional but highly recommended if you need 100% accuracy or you have no other way of double-checking with another source of truth.

- **Data extraction training**: This sub-step passes the human-validated information back to the classifiers to improve their future predictions. This sub-step closes the feedback loop for extractors that are capable of learning from human feedback. The only available extractor trainer is the Machine Learning Classifier Trainer.

5. **Export**: This step allows you to export validated data for consumption in the automation workflow. The validated data is exported as a DataSet variable. The variable can be saved into a DataTable format and be consumed in other systems.

As we have seen in this section, **UiPath Document Understanding** is a powerful product that leverages AI to enable an RPA robot to understand documents. Let's now look at UiPath AI Center.

Getting to know UiPath AI Center

UiPath AI Center is an important product to familiarize yourself with because it allows you to tackle a new set of use cases by incorporating AI and **ML** models into your RPA automations with the help of UiPath AI Center. UiPath AI Center is an ML model life cycle management tool and is designed to make it easy for RPA developers to build, manage, and effectively govern ML models. It sits in the Manage pillar of the UiPath platform.

The benefits of UiPath AI Center

The following diagram shows the benefits of using UiPath AI Center in the UiPath platform:

Figure 3.14 – UiPath AI Center: Benefits

UiPath AI Center provides six key benefits:

- *OOTB models from UiPath and partners*:

    UiPath AI Center has over 25+ OOTB models from UiPath and many more from the wide network of UiPath partners. These OOTB models cover UiPath Document Understanding, Language Analysis and Comprehension, Image Analysis, and Tabular Data. RPA developers without deep ML backgrounds can start using these OOTB models to incorporate AI into their use cases.

- *Bring your own models*:

  UiPath AI Center supports users bringing their own ML models into UiPath AI Center to deploy, continuously train, and improve these models. This feature allows RPA developers to collaborate with data scientists in their companies to use the models the data scientists incorporated into their RPA workflows.

- *Human validations*:

  UiPath AI Center allows humans to work alongside machines to validate data and send it back for model retraining. To maximize automation potential, features that help with human-machine collaboration are essential.

- *Continuous improvement and re-training*:

  UiPath AI Center enables end-to-end visibility on model use, data, model performance, user actions, and pipelines. It also maintains control over versions and retrains without strain with the human in the loop and automatic retraining.

- *Multiple deployment options*:

  UiPath AI Center allows for on-premises deployment and cloud deployment. These options satisfy the requirements of all industries.

- *Intuitive and secure data management*:

  UiPath AI Center has an intuitive, yet secure, data management feature. It is simple to create and access data for the models. It also allows users to upload their own data. In addition, it enables customers to snapshot the data that they trained their model on so that they have a full audit trace of which model made which prediction and which data was trained on which model.

UiPath AI Center technical concepts

Let's walk through the technical concepts of **UiPath AI Center**. You will have a good overview of each concept and how they relate to each other. The following diagram shows the technical concepts of UiPath AI Center:

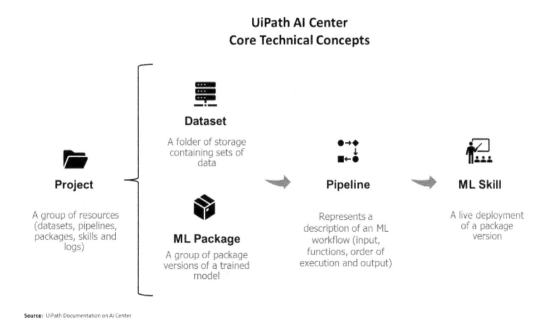

Figure 3.15 – UiPath AI Center: Core technical concepts

UiPath AI Center uses five technical concepts. You need to be familiar with the concepts and how they relate to each other. They are used throughout the UiPath AI Center product. The five technical concepts are as follows:

- **Project**: A project is a collection of resources – datasets, ML packages, pipeline ML skills, and logs. You use these resources to build a specific ML solution. It is best to create a separate project for each automation use case. Within **UiPath AI Center**, the dashboard provides an overview of all the projects with information on active pipelines and deployed packages.

- **Dataset**: A dataset is a folder of data storage. You can create new data files or upload files from another application. A dataset is a local resource. It is available only within the project.

- **ML Package**: An ML package is a folder with the code and metadata needed to train an ML model. An ML package can have multiple versions with associated change logs. In UiPath AI Center, there are 25+ OOTB ML packages to use.

- **Pipeline**: A pipeline is the linear sequence of steps required to prepare the data, tune the model, and transform the predictions. A pipeline specifies the functions and the order of executions of these functions. It also specifies inputs required to run the pipeline and outputs received from it. Once a pipeline is completed, a pipeline run has associated outputs and logs. There are three types of pipelines:

  - **Training pipeline**: A training pipeline is used to train a new ML model to produce a new package version.

  - **Evaluation pipeline**: An evaluation pipeline is used to evaluate a trained ML model.

  - **Full pipeline**: A full pipeline is used to train a new ML model and evaluate the performance of this new model. Additionally, a preprocessing step is run before training, allowing data manipulation and training of a trained ML model.

- **ML Skill**: An ML skill is a consumer-ready, live deployment of an ML package. Once deployed, it is ready to be consumed within RPA workflows. In UiPath Studio, you simply drag and drop any ML skill activity into the RPA workflows.

As you have seen in this section, UiPath AI Center allows you to tackle a new set of use cases by incorporating AI and ML models into your RPA automations with the help of UiPath AI Center. Let's look at how the UiPath platform handles intelligent chatbot capability with the UiPath chatbot with Druid.

Getting to know the UiPath chatbot with Druid

The **UiPath platform** and intelligent chatbots in the market bring together RPA and conversational AI to enhance your customer and employee experience.

The UiPath platform is flexible and offers several options for integrating intelligent chatbots. UiPath has partnered with the leading chatbot providers to provide direct integrations. UiPath also has a built-in connector for the Dialogflow platform. Lastly, UiPath has natively integrated the no-code, AI-powered chatbot from Druid into the UiPath platform. Let's dive into the Druid chatbot as it is part of the UiPath platform.

Benefits of the UiPath chatbot with Druid

The following diagram shows the benefits of using the UiPath chatbot with Druid in the UiPath platform:

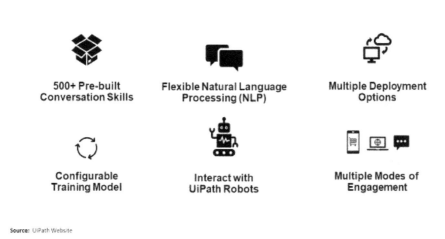

Figure 3.16 – UiPath chatbot with Druid: Benefits

The UiPath chatbot with Druid has six key benefits:

- *500+ pre-built conversation skills*:

  The UiPath chatbot with Druid has over 500 pre-built skills for business scenarios across multiple industries and roles. These conversation skills include replying to questions, sending notifications, delivering reports, tracking smart tasks, filling in forms, and routing to humans. Druid is a no-code authoring platform that allows RPA developers to create and deploy chatbots quickly without an in-depth technical knowledge of AI.

- *Configurable training model*:

  The UiPath chatbot with Druid is equipped with comprehensive training tools and learning mechanisms to improve the knowledge and performance of the chatbot.

- *Flexible NLP*:

    The UiPath chatbot with Druid has a proprietary NLP technology but also allows users to pick other NLP technologies. The NLP engine interprets the user's intent to provide information contextually based on their behavior and preferences. The platform also has advanced NLP features to test utterances, configure stop words, sentiment, and flow matching thresholds.

- *Interact with UiPath robots*:

    The UiPath chatbot with Druid can perform bidirectional communication between robots and humans using natural language to process requests. The following diagram shows how the integration between UiPath and the Druid chatbot works:

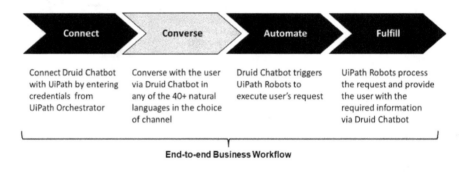

Figure 3.17 – UiPath chatbot with Druid: Integration

UiPath and Druid have native dynamic integration to incorporate the Druid chatbot into the RPA workflows.

- *Multiple deployment options*:

    The UiPath chatbot with Druid can be deployed on-premises, in the cloud, and as a hybrid. The options fulfill the requirements of different industries and customers.

- *Multiple modes of engagement*:

    The UiPath chatbot with Druid supports ready-made integrations for omnichannel deployments through web pages, collaboration tools such as Microsoft Teams or Slack, chat applications such as WhatsApp or Facebook, or voice calls.

Technical components of the Uipath chatbot with Druid

Let's walk through the technical components of **the UiPath chatbot with Druid**. You will have a good overview of each component. You will understand how best to incorporate Druid chatbot in your next project. The following diagram shows the technical components of the UiPath chatbot with Druid:

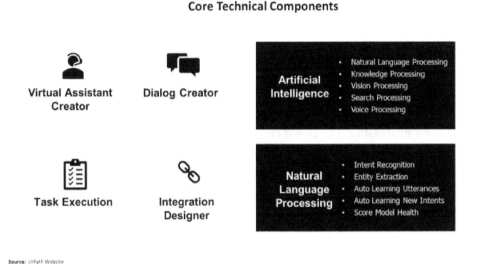

Figure 3.18 – UiPath chatbot with Druid: Core technical components

The UiPath chatbot with Druid includes six technical components. They are as follows:

- **Virtual Assistant Creator** is a graphical flow designer. It enables RPA developers to create, design, and edit conversational flows. There are over 500 pre-configured conversational skills and pre-built virtual assistant templates to use.

- **Dialog Creator** is a drag-and-drop tool. The tool has built-in, real-time conversational analysis to enhance future conversations. It supports over 40 languages OOTB. It also comes with pre-built conversational templates for common business workflows customized for roles, processes, and industries. You can also implement your own static responses and behavior.

- **Task Execution** can monitor and alert users in real time during conversational streams. The Druid chatbot allows users to give tasks to other users and monitor their completion. It can also be trained to send alerts and follow-ups if tasks were not executed in the predefined time interval.

- **Integration Designer** is where the configuration of enterprise applications and open APIs takes place. The Druid chatbot can quickly connect with other enterprise systems – CRM, ERP, BI, HRIS through REST, SOAP, web services, APIs, MQs, and SQL connection. Predefined connectors with the most common IT systems are available. In addition, the integration with the UiPath platform is native. Druid chatbot features are natively integrated with UiPath Studio, UiPath Orchestrator, and attended robots to support triggering automation directly from the chat requests.

- **Artificial Intelligence**: The Druid chatbot is AI-enabled. Its AI capabilities include the following:

  - **Natural Language Processing**: The Druid chatbot can identify intent, extract context, and acknowledge the sentiment.

  - **Knowledge Processing**: The Druid chatbot can map complex information and data to provide intelligent recommendations and perform a semantic search.

  - **Vision Processing**: The Druid chatbot can process images and identify and capture critical information with a high level of accuracy.

  - **Search Processing**: The Druid chatbot can use search APIs to comb web pages, images, videos, and news with a single API call.

  - **Voice Processing**: The Druid chatbot can convert spoken audio into text, use voice for verification, or add speaker recognition.

- **Natural Language Processing**: **The Druid chatbot** allows users to pick their preferred NLP engine. With the help of the selected NLP engine, the Druid chatbot can interpret the user's intent to provide contextual information based on the user's behavior and preferences. The NLP engine has the following components:

  - **Intent Recognition**: Match the utterance with its correctly intended task.

  - **Auto Learning Utterances**: Automated utterance generation to improve precision and recall for retrieving the right answer.

  - **Auto Learning New Intents**: Allow unsupervised learning by automatically adding the successfully executed utterances to its ML training set.

  - **Score Model Health**: Validate the performance of the model using the confusion matrix and overlapping utterances.

UiPath has natively integrated the no-code, AI-powered chatbot from Druid into the UiPath platform. The UiPath platform with Druid has many benefits. Druid has over 500 pre-built conversation skills, configurable NLP and training models, multiple deployment options, multiple modes of engagement, and native integration with UiPath.

Summary

In this chapter, we explored all aspects of the UiPath platform. The UiPath platform is ideal for companies that are looking to accelerate and amplify their transformation journey with cognitive automation. We have learned the critical success criteria for a cognitive automation platform. We provided an overview of the entire UiPath platform. We also provided an overview of three key AI-infused products – UiPath Document Understanding, UiPath AI Center, and UiPath Chatbot with Druid.

What you learned in this chapter will serve as a foundation for you to get hands-on experience with the examples involving these products in the subsequent chapters.

In the next chapter, we will explore how to identify cognitive opportunities. As we work to identify automation opportunities, the value proposition of automation should always be the central focus, tying back to the higher-level targeted business objectives of the organization and the central vision of the automation program.

Section 2: The Development Life Cycle with AI Center and Document Understanding

In this section, we will learn how to integrate AI Center and Document Understanding in automation development opportunities by following the traditional development life cycle.

This section comprises the following chapters:

4
Identifying Cognitive Opportunities

As more organizations implement **Robotic Process Automation (RPA)** and cognitive automation within the workplace, generating value from these implementations has never been more important. As we work to identify automation opportunities, the value proposition of automation should always be the central focus, tying back to the higher-level targeted business objectives of the organization and the central vision of the automation program.

When identifying potential use cases for cognitive automation, it is imperative to gather as much context and knowledge about an automation opportunity to ensure AI Center and Document Understanding can achieve the use case's target goals. Therefore, in this chapter, first, we will focus on how to search for automation opportunities. In particular, we will learn about the characteristics of a traditional automation opportunity and a cognitive automation opportunity. How to set target goals for the enterprise, and how to lead search efforts to find automation opportunities. Following this, we will dive into understanding the automation opportunity: learning how to prioritize automation opportunities based on the dimensions of best fit, feasibility, and viability and learning how to best capture value by fully understanding the end-to-end process of an automation opportunity. By the end of the chapter, we'll pivot into discussing what items to consider when probing a cognitive automation idea.

In this chapter, we're going to cover the following main topics:

- Searching for automation opportunities
- Understanding the opportunity
- Probing for cognitive automation

Searching for automation opportunities

Value assurance has never been more important, as organizations are laser-focused on creating immediate benefits from automation initiatives. Therefore, careful consideration should be made when identifying, sizing, and green-lighting automation candidates during the opportunity search. To ensure that the chosen opportunities are successful and realize the benefits advertised, you should be well versed in the characteristics of an automation opportunity, understand when to apply cognitive capabilities, and understand how to recognize the drivers of value automation can create. Once fully versed in the characteristics of an automation opportunity, you can take different approaches to identify an opportunity from a top-down or bottom-up perspective, and then apply your judgment to prioritize the opportunity for further analysis and development.

In this section, we're going to cover the following topics:

- The characteristics of an automation opportunity
- Identifying target goals
- Seeking automation opportunities

The characteristics of an automation opportunity

Before searching for automation opportunities within the enterprise, it is important to fully understand what traditional RPA can do, and more specifically, what additional benefits adding cognitive abilities to RPA can unlock. Fully grasping these concepts is pivotal when searching for opportunities – far too often, poorly scoped opportunities end up within an organization's automation pipeline that are either not well fit for automation or unable to provide immediate benefit to outweigh the cost of development and support.

In previous chapters, we gained a general understanding of what RPA is and what it can offer – a software robot residing on top of applications at the desktop level that can mimic human interaction to perform a series of steps. This automation can provide a lot of benefits, such as the following:

- Logging into applications

- Scraping data from websites and applications

- Navigating through application windows and menus

- Making calculations

So, when identifying candidates for strictly RPA, we want to investigate opportunities that do the following:

- Require manual interaction with applications

- Are repetitive in nature

- Have standard inputs

- Are time-consuming

- Are rules-based and structured

However, it's critical to understand that RPA alone can only perform a series of standardized, rules-driven system-based activities. In its basic form, RPA is very naïve, only following the instructions that it's provided. Of course, this is similar to any kind of traditional programming or scripting – your executing agent (in our case, the UiPath robot) is simply going to follow the set of instructions you provide it. Unfortunately, this leaves us with a finite number of opportunities that are automatable, leaving out a lot of use cases that require a human-level form of judgment or thinking outside of the traditional if-else series of conditionals.

Fortunately for us, we can now fill this gap with the addition of cognitive abilities to what I like to call "cognitive automation." This allows us to potentially reshape the future of automation, as we can scope more complex and judgmental types of opportunities that traditional rules-based automation couldn't solve. Many organizations are leveraging this technology to improve their existing automations and to even investigate their automation pipelines and revisit previously descoped opportunities that can now be developed with cognitive automation.

Now, just because we can include cognitive abilities to unearth additional automation opportunities, it's important to realize that cognitive automation can't solve every problem or, at least, not easily. Therefore, the following characteristics of an automation opportunity necessitate the need for a cognitive approach:

- **Probabilistic in nature**: This describes a use case where the determination of a result is not certain. Examples of use cases that are probabilistic in nature include *property valuations* or *probability of default*, where we can't determine an outcome with complete certainty.

- **High variability**: This describes a use case that contains many deterministic conditions (*if this… then that*). Examples include *resume classification* or *language understanding*, where there is a large amount of variability that can potentially be inferred with statistical models.

- **Unstructured data**: This describes a use case where data is not standardized. Examples include *email classification* or *invoice extraction*, where the input data does not conform to a regular schema.

Now that we understand the characteristics of traditional RPA and cognitive automation, it is also important to understand the types of value that automation can provide. We want to target automation candidates that are not only feasible to be developed but also provide impact to our end users. With this understanding, we can search the enterprise for potential automations that can be confidently developed and can provide the value the business expects. The effects of implementing RPA are widely known, and increasingly, more organizations are looking to automation to provide impact, such as the following:

- Increasing capacity
- Improving quality
- Growing revenue
- Improving cycle time
- Enhancing experience

In conclusion, when reviewing and scoping candidates for automation, ensure that the candidate conforms to the characteristics of RPA (that is, requires manual interactions, is repetitive in nature, has standard inputs, is time-consuming, and is rules-based). Following these characteristics should ensure that the candidates chosen for automation are the best fit to be automated. However, if there exists an automation candidate that is not completely rules-based, then Cognitive automation can be considered for the opportunity if the opportunity is either probabilistic in nature, highly variable, or contains unstructured data. In addition to considering cognitive abilities, candidates that are not completely rules-based can also be considered for alternative types of automation, such as attended or human-in-the-loop types of automations, where automations and humans pass information between each other. Finally, if the candidate falls into the preceding characteristics, we want to investigate whether value can be derived from the opportunity, such as the amount of time saved or the capacity freed.

Identifying target goals

As we start to seek automation opportunities, it is imperative that you align the organization and its automation program to a series of business objectives. These business objectives will better align the automation program with a set of criteria to look for when investigating automation opportunities. Following a systemic, outcome-driven approach, as shown here, will keep us focused on creating value for the enterprise (*Figure 4.1*):

Figure 4.1 – An outcome-driven approach for creating value

For many new automation programs, it might be compelling to automate low-complexity automations. However, by finding automation opportunities that align with a strategic priority, we can make more compelling value propositions and more compelling success stories when trying to rally engagement around automation. Therefore, the outcome-driven approach starts with leadership, setting a business objective, and forming a hypothesis (backed by data) to identify key pain points. Typical business objectives of automation programs include the following:

- Growing revenue
- Improving client experience
- Scaling faster
- Improving compliance/regulations
- Reducing cost
- Improving employee experience
- Building a foundation for AI

Once the business objectives are chosen, an investigation into the organization can lead us to create hypotheses that can be potentially resolved with automation. For example, if an organization wants to focus on improving employee experience, first, we need to dive into what impacts the experience of an employee – ideally, job satisfaction and support (for example, feeling valued).

Knowing what factors influence our business objective, next, we can focus on data collection and the investigation into the enterprise. For this example, we are looking at employee satisfaction, so we should be collecting any data HR has related to satisfaction surveys, support tickets, and more (*Figure 4.2*):

Survey Results		
①	**68%**	Found the onboarding process inhibiting productivity during an employee's first months
②	**60%**	Do not see their work as rewarding
③	**35%**	Are unable to see professional growth within their position

④	Key Performance Indicators		
	KPI	**CURRENT**	**TARGET**
	Employee Satisfaction	70%	85%
	Accuracy	95%	97%
	Timeliness	65%	80%

Figure 4.2 – An example of data collection results

From the data collection, we can start to create hypotheses:

1. A majority of respondents found onboarding to be inhibiting productivity.

   Hypothesis: Manual setup across multiple applications and systems leads to slower onboarding times and reduced productivity.

2. 60% of respondents do not find their work rewarding.

   Hypothesis: Business units contain a lot of manual and tedious tasks that small teams must complete, thus reducing the amount of time that employees spend performing rewarding tasks.

3. 35% of respondents cannot see professional growth.

   Hypothesis: Employees do not have a prescribed career trajectory.

4. The service center has a poor turnaround time.

   Hypothesis: Many applications and slow data retrieval lead to increased turnaround times.

And from these hypotheses, we can start to pinpoint areas that are necessary for further investigation. For this example, let's investigate hypothesis 1 and venture into the onboarding journey of a new employee. In this case, we would want to dive into the employee life cycle, reviewing the journey from applying for roles, interviewing, accepting the offer, and onboarding. In order to do that, we would set up discovery meetings with respective stakeholders from the teams responsible for performing those activities. In the next section, we'll look, in more detail, at how to run these types of discovery meetings.

Before we jump into the different approaches for seeking opportunities to automate, there are a number of success factors to consider:

- Keep the targeted business outcome at the forefront of ideation.
- Ensure executive alignment and buy-in to maximize impact.
- Avoid exclusively focusing on cost reduction.
- Focus on solving business problems.
- Consider other types of methods and solutions – RPA is just one tool in the toolbox.

Following these success factors can help ensure we identify automation opportunities that provide value to the business and the automation program as a whole.

Seeking automation opportunities

As we seek automation opportunities throughout the enterprise, we will notice different levels of automation, including levels going from *Bottom-Up* to *Top-Down*. These types of automation can be referenced in *Figure 4.3*:

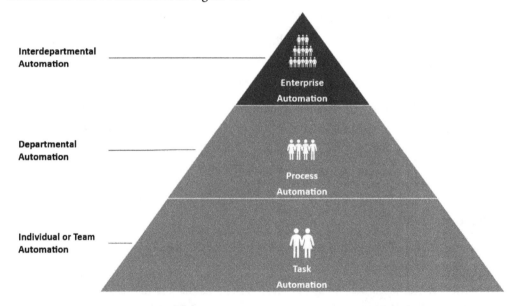

Figure 4.3 – The types of automation opportunities

- **Enterprise automation** looks at interdepartmental opportunities that impact a large population of the organization. These automations might include higher complexity technologies such as AI.

- **Process automation** looks at departmental opportunities that are uniquely developed for a department's set of procedures. These opportunities might provide high levels of impact with multiple colleagues confined within a department. Automations could include additional technologies such as document understanding, AI, or "human in the loop."

- **Task automation** looks at low-complexity opportunities that are for the benefit of a limited number of individuals. These automations are built to automate the unique tasks an individual performs on a day-to-day basis. In general, these opportunities provide a lower impact on automation programs but can lead to a larger impact on the individual who is running the automation. These automations generally do not include higher complexity technologies.

As great as enterprise automation is (in terms of impact), successful automation programs employ both a top-down and bottom-up approach for pinpointing automation opportunities. Not all automations will be at a company-wide level; therefore, it is imperative to also search for departmental-level and task-level automation opportunities. Let's investigate a few approaches of top-down and bottom-up ideation (please refer to *Table 4.1* and *Table 4.2*):

Top-Down Ideation	
	CoE Discovery Workshops
Approach description	• Smaller, more focused intake effort where the **Center of Excellence (CoE)** targets specific functional areas (of a business) to identify discrete automation opportunities in order to answer a general business objective. This effort happens in discrete workshops, qualifying ~<10 opportunities at a time.
Key business responsibilities	• Identify key resources to support ideation efforts and participate in education sessions. • Survey opportunities across the business for review with CoE.
Key CoE responsibilities	• Conduct RPA Education and guide businesses to identify opportunities. • Facilitate intake workshops and meetings to qualify the business opportunities identified.
Key activities	• Hold RPA education sessions (as needed) and understand the business objectives for leveraging automation. • Business to investigate opportunities within desired space (or engage CoE to help ideation efforts). • CoE to hold intake workshops/meetings to qualify opportunities.

Table 4.1 – The top-down approach; CoE discovery workshops

Top-down automations are usually tied to larger enterprise-wide digital transformation initiatives. Because these initiatives are well-funded programs that focus on the strategic objectives of the company, it's usually easier to get interdepartmental cooperation to participate in ideation exercises. With this cooperation, one of the best ways to identify opportunities from a top-down perspective is to submit data collection requests, asking for function organization charts (such as sub-functions/how teams are organized, headcount, and more). This data will allow us to focus on areas of high headcount that we can probe into.

These identified areas can then be focused on with smaller intake efforts, where we can identify discrete opportunities for automation in order to answer a general business objective. Once a general business objective has been established (such as speeding up tasks for an accounting team), we can investigate the processes performed and qualify opportunities for automation:

Bottom-Up Ideation	
Business Ideation and Submission	
Approach description	• Larger, exhaustive data collection effort where the CoE works with multiple functional areas to cast a net across the enterprise. This effort leverages multiple workshops/education sessions to build a pipeline of initiatives.
Key business responsibilities	• Define an automation strategy and long-term plan. • Attend automation education sessions. • Submit ideas into Automation Hub for CoE engagement.
Key CoE responsibilities	• Ensure that operating model control processes are being followed. • Work with the business to forecast demand and plan for building opportunities.
Key activities	• CoE invites business leaders to education session(s) to share objectives and introduce RPA and Automation Hub. • CoE collects automation opportunities in Automation Hub from business leaders and participants across the organization. • CoE reviews opportunities and identifies use cases for further analysis and building.

Table 4.2 – The bottom-up approach; business ideation and submission

From a bottom-up perspective, the approach is somewhat flipped from the top-down approach. With bottom-up ideation, ideas need to come from the organization versus the targeted approach of top-down.

For bottom-up, we perform larger data collection efforts where the entire organization (or a subset) is given the opportunity to submit automation ideas that they believe are impactful. Participants are welcomed to education sessions where they can share the objective(s) of automation and the art of the possible. Once educated, participants are welcomed to submit potential automation ideas into a central repository (such as Automation Hub), where each idea is reviewed for further analysis and building.

Once we've completed our exercises to probe for opportunities, our **Master Opportunity Log** should contain a number of opportunities for us to deep-dive into. This master opportunity log should be an organized list based on the following high-level information that we might have gathered from our ideation exercises, such as the following:

- Value
- Complexity (a rough estimate)
- Suitability (a rough estimate)

This master opportunity log can take many forms – some organizations choose a tried-and-true Excel sheet to keep track of the pipeline, but many organizations leverage tools such as UiPath's Automation Hub or Jira to keep track of an automation pipeline:

Automation Pipeline

All 4	Review 0	Decision Pipeline 1	Implementation 1	Live 2

Q Search...   = Sort ˅   ⊞ Columns ˅

☐ Automation Name	Categories ⇅	Submitter's Department ⇅	Phase ⇅
☐ **My great automation idea** Last Modified: Date Submitted:	Finance & Accounting ⌐ Accounts Receivable - Order to Bid ⌐ Opportunity Identification	Finance	Qualification
☐ **Test Cit Automation** Last Modified: Date Submitted:	Finance & Accounting ⌐ Accounts Receivable - Order to Bid ⌐ Opportunity Identification	Finance	Live
☐ **Gather Acme-Test Account Details** Last Modified: Date Submitted:	Finance & Accounting ⌐ Record to Report - General Accounting ⌐ Master data management (including ch	Dept	Live
☐ **CoE Driven Idea** ➍ Last Modified: Date Submitted:	Finance & Accounting ⌐ Accounts Receivable - Award to Contract ⌐ Pricing Review	Finance	Analysis

Figure 4.4 – The master opportunity log in UiPath's Automation Hub

From this simple prioritization, we can now double-click on each opportunity and further assess which projects our team can start diving into and developing.

In this section, we reviewed how to ensure a value-based approach when searching for automation opportunities. Successful automation implementations rely on providing value to the business. We learned about the characteristics of an automation opportunity, how to identify target goals, and how to search the business for automation. As we identify potential automation opportunities, we need to dive deeper into each opportunity before starting implementation work. In the next section, we will look at how to understand potential automation opportunities.

Understanding the opportunity

Building and maintaining an automation pipeline is the catalyst to a successful automation program. A successful automation pipeline should always be flowing with ideas of differing levels of complexity and value. In this section, we will investigate how to keep the automation pipeline staffed by evaluating, investigating, and prioritizing opportunities.

Evaluating opportunities

Now that our master opportunity log is full of automation opportunities and is prioritized based on estimated data (including value, complexity, and suitability), we can venture into each opportunity and assess which candidates are best suited for automation. During these assessments, additional meetings with business stakeholder(s) might be conducted to vet the feasibility and calibrate the qualitative ratings before starting implementation. For vetting and further evaluating the opportunities within our master opportunity log, we perform the following exercises:

- A best-fit assessment (is the process the best fit for RPA?)
- A feasibility assessment (is the process feasible to automate?)
- A viability assessment (what value does the process generate?)

Best-fit assessment

Not all opportunities are the best fit for sole automation; therefore, the **Best-Fit Assessment** is an initial exercise to test the suitability of an automation opportunity. Before diving into the details of feasibility and impact potential, the opportunity should be reviewed to determine whether RPA or another technology is the best fit for the opportunity.

As we start assessing automation opportunities, we can leverage the following matrix to easily determine RPA suitability:

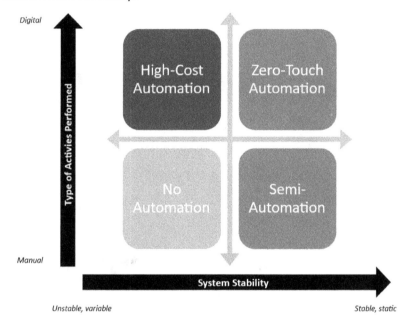

Figure 4.5 – Automation opportunities can be plotted within one of the four quadrants

- **No automation**: Generally, we want to be cautious of opportunities that involve frequent process and environment changes (for example, volatile system environments and frequent procedural changes). Thus, when assessing opportunities, we want to stay away from processes that are highly physical and contain volatile processes or systems.

- **High-cost automation**: As opportunities drift toward being more digital, we start to cover use cases that are exclusively performed on computers. Consideration of the volatility of systems and environments is still necessary. However, process changes might require the integration of new software components or tools – in many cases, cognitive computing is necessary, such as image recognition.

- **Semi-automation**: As we drift downward into more static processes and environments, opportunities that contain more physical steps can potentially be semi-automated. While RPA cannot cover the physical steps, there is still an opportunity to rescope or reengineer the opportunity for automation.

- **Zero-touch automation**: Lastly, the best use case opportunities are fully digital and contain stable systems/environments. These are well suited for automation even if we need to include cognitive components.

During the best-fit assessment, you should sit with the process owner or a **Subject-Matter Expert (SME)** of the process to probe into the details of the as-is process. By learning more about the process and how it's performed, referencing a checklist such as in *Figure 4.6* can help you decide whether the opportunity is fit for automation:

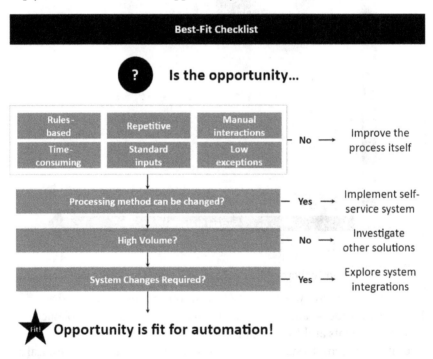

Figure 4.6 – An example best-fit checklist

When reviewing an automation opportunity for best fit, first, we must review whether the opportunity contains the characteristics of typical automation (that is, requires manual interactions, is repetitive in nature, has standard inputs, is time-consuming, and is rules-based) or of cognitive automation (that is, probabilistic in nature, high variability, and unstructured data). If the opportunity does not conform to these characteristics, then the opportunity likely needs to undergo some process improvement exercises to make the process fit for automation.

If the opportunity conforms to the characteristics of typical or cognitive automation, then we must next check to see whether the processing method can stay unchanged. If we find that the transactions of a process could be handled by an application (for example, using ServiceNow to handle service tickets instead of RPA), then the opportunity might be better fit for self-service than automation.

If we can continue, then we need to ensure that the opportunity has enough volume to support the building of the automation. This will tie in with the viability assessment, but generally, we want to look at automation opportunities that are performed on a regular basis (daily, monthly, or even quarterly) that contain a high number of transactions. If the volume is low, the opportunity may be a better fit for another solution outside of automation.

Finally, if we can continue, we need to check whether any system changes would be required for automation. For automation, we want to ensure application and system stability, so we want to check for the following:

- The applications or systems aren't changing soon.

- There is already functionality (or upcoming functionality) within the existing applications and systems that would be best for the opportunity.

In addition to the example best-fit checklist shown earlier, UiPath's Automation Hub includes an automation potential score to indicate the best fit (*Figure 4.7*):

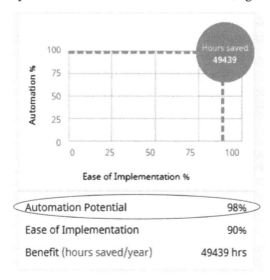

Figure 4.7 – The automation potential score within Automation Hub

The automation potential score within UiPath Automation Hub will output a percentage based on the following:

- How structured the input data is

- Process variability

- How digitized the input data is

By completing the best-fit assessment, we can gauge the potential of fitting an opportunity with automation. With the best-fit assessment completed, we can move on to gauge how difficult implementing automation would be with the feasibility assessment.

Feasibility assessment

Once RPA is determined as the best fit for the opportunity, the next assessment should be determining the ease of implementation, known as the **Feasibility Assessment**.

When checking for feasibility, we want to check several factors that can raise the complexity of development efforts:

- **The number of screens**: RPA works at the application layer, so with every new screen (or window) added to a process, a developer would need to configure automation to interact with elements on each page. The number of screens can also be taken as a proxy for the number of steps.

- **The type of application**: The application type is a large contributor to complexity. Applications such as Java or Citrix might increase the complexity of development, while others such as Mainframe, Web, or Microsoft-based applications might decrease complexity. Another point to consider is not only the type of application but the number of applications too, as more applications in a process increase the amount of configuration a developer would need to develop.

- **Variations/scenarios/exceptions**: Any variation in the process (for instance, the if-else types of rules) will lead to an increase in complexity, as more development time will be needed to handle each variation.

- **Structured inputs**: Input type is another factor affecting complexity. Input type can vary between machine-readable inputs (text) and digital inputs (scanned PDFs), leading to different methods of reading text. In addition to the input type, the structure of the input (for example, standardized, semi-structured, or unstructured) also contributes to complexity. An automation that needs to interpret free text from an email body will be more complex to develop than an automation that searches for static email subject lines.

- **Quality control**: Certain use cases might require a high level of accuracy and/or **strict service-level agreements (SLAs)**. While automation can improve accuracy by always having the same series of steps, processes that require more complex components (such as **Optical Character Recognition** or **OCR**) might be more difficult to maintain at a high level of accuracy.

- **Image-based automation**: Automation that must interact with VDIs, Remote Desktop, or Citrix-based applications can increase development complexity. While UiPath's Remote Desktop and Citrix extensions have made the development of remote desktops easier, some components of development could require image-based automation that can increase development and testing times.

- **Regulatory compliance**: Many process opportunities come with regulatory obligations, especially from a banking or healthcare space. Regulatory compliance might require additional scope meetings, approvals, and audits to ensure automation is compliant.

The responses from the preceding factors can be combined to get a rough estimate of process complexity. The following table (*Table 4.3*) can be referenced to show how process characteristics map to process complexity:

Process Complexity	Low	Medium	High
# of Applications	0 - 3	3 - 5	5 - 7
# of Fields	1 - 99	100 - 200	201 – 300
# of Process Steps	1 – 20	20 - 60	60+
# of Screens	0 - 10	11 - 20	21 – 30
# of Variations	0 - 3	4 - 6	7 – 8
# of Exceptions	1 – 10	10 – 20	20+
Process SLAs?	No	Yes	Yes
Image-Based Automation?	No	No	VDI/Remote Desktops
Number of Input Formats	2 - 3	3 - 5	5 - 7

(Row label on left side: Process Characteristics (Non-Exhaustive))

Table 4.3 – The process complexity matrix

By mapping all of the process characteristics that increase complexity within a matrix, we can effectively run a high-level estimate on the complexity of an automation opportunity by fitting the details of the opportunity into the matrix. Then, by assigning timelines to each process complexity (for example, low complexity taking 3–4 weeks), we can also run a high-level estimate of how long an implementation might take.

Similar to the best-fit assessment shared earlier, UiPath's Automation Hub includes an ease of implementation score to indicate feasibility (*Figure 4.8*):

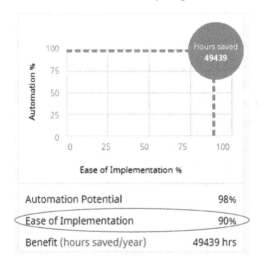

Figure 4.8 – The Ease of Implementation score within Automation Hub

The ease of implementation score within UiPath Automation Hub will output a percentage based on the following:

- Process stability
- Application stability
- How structured the input data is
- Process variability
- Process length
- The number of applications used

By completing the feasibility assessment, we can gauge how difficult implementing automation would be to a particular process. If the automation opportunity passes the feasibility assessment, we then need to review the potential impact of the opportunity to justify the development cost with the viability assessment.

Viability assessment

After determining the best fit and feasibility of the opportunity, a business case should be created to justify the value or impact of implementing automation. There are many ways to capture an automation's value. The following list of benefits are examples of typical automation ROI:

- **Cost savings**: Cost reduction from **Full-Time Equivalent** (FTE) savings

- **Productivity gain**: Improvement in turnaround time or an increase in the number of volumes handled.

- **Error reduction**: A reduction in error due to the automated nature versus the tendency for errors from manual human interaction

- **Client satisfaction**: The increase in client (internal or external) satisfaction due to automation performance and output

- **Flexibility**: The ability to scale and descale during changes in volume

- **Business agility**: The indirect impact on downstream business units to act/respond at a quicker pace due to improved turnaround time by automation

- **Compliance**: The increased ability to comply with regulatory requirements (such as the ability to fully audit automated transactions or the ability to restrict automation access)

Ideally, the benefits derived from an automation opportunity should align with the target goals we discussed earlier, so an analyst should take all considerations into account when ranking an opportunity. While more emphasis is generally placed on the qualitative measures of impact (such as cost savings and productivity gain), softer types of impact (such as satisfaction and error reduction) can also be valuable measures of automation impact.

Similar to the best-fit and feasibility assessments, UiPath's Automation Hub (*Figure 4.9*) provides a benefit score to measure viability:

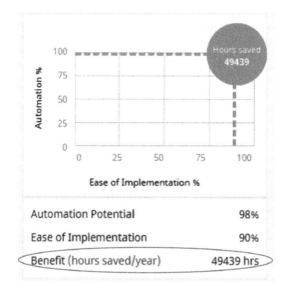

Figure 4.9 – The Benefit score within Automation Hub

If you are trying to capture time savings, the benefit score within Automation Hub is a great way to estimate the potential benefit in hours saved per year.

By completing the best-fit, feasibility, and viability assessments, we should be able to trim our backlog of automation opportunities into those most probable to provide value and impact to the organization. Now that our opportunities are vetted, we can continue with prioritizing the remaining opportunities within our pipeline.

Prioritizing the pipeline

After an opportunity has gone through the evaluation of best fit, feasibility, and viability, the opportunity is now ready to be placed into the **Use Case Backlog**. This backlog consists of all the opportunities that have been vetted and are ready for development efforts.

In order to prioritize this backlog, we can map each opportunity based on its complexity and its value, all derived from the assessment exercises we performed earlier. To help better compare opportunities, we can create a 2 x 2 matrix, with complexity as our *y* axis and benefits as our *x* axis:

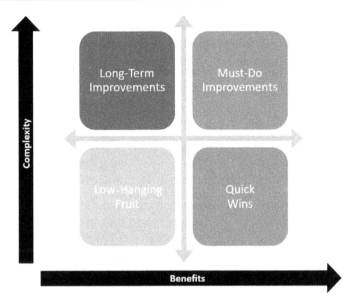

Figure 4.10 – Mapping automation opportunities into quadrants

Within this matrix, we will map opportunities into one of four quadrants:

- **Long-term improvements**: These are higher complexity automations that provide little to no benefit. Generally, try to avoid these types of projects, and probe into other types of improvements such as system/application upgrades, process improvements, then automation.

- **Must-do improvements**: These are opportunities that provide greater benefits but are more complex to develop. Most cognitive use cases would reside within this bucket. They are best to tackle must-do improvements once all the quick wins have been exhausted.

- **Low-hanging fruit**: These are opportunities that provide fewer benefits but are easier to develop. They are best when needed to show the art of the possible, but should not be tackled if there are quick wins available.

- **Quick wins**: These are the perfect type of opportunity; they should be at the top of any pipeline. Quick wins provide quick impact at a low cost, and all quick wins should be exhausted before moving on to the must-do improvements or low-hanging fruit.

From these four quadrants, we can start to create an implementation plan – strategizing how we plan to build the automations within our pipeline. By taking a phased approach, we can split our implementation plan into distinct waves, aligning each of the quadrants into a distinct wave:

- **Wave 1**: The first wave of our implementation plan should be focused on showcasing value and impact. Therefore, we should be selecting the use cases identified as *quick wins* and *low-hanging fruit* first. These use cases should be lower complexity with notable impact – not only showcasing the potential viability of automation but also getting a broader buy-in from stakeholders.

- **Wave 2**: During the second wave, we want to continue showcasing the impact of the first wave. We should keep our approach centered on completing all the *quick wins* and *low-hanging fruit* within the pipeline. By the end of the second wave, confidence and excitement around automation should be building. It is not uncommon for the pipeline to continue organically growing as more business units see the success of automation.

- **Wave 3**: By the third wave, our original list of *quick wins* and *low-hanging fruit* should be exhausted. By this point, our focus can be shifted to the other quadrants, *must-do improvements* and *long-term improvements*. Additionally, it is not uncommon to take on more complex style use cases during this wave. Building on the value-generation theme of waves one and two, wave three can be focused on showcasing the "art of the possible" by venturing out into more complex use cases, such as the cognitive style use cases, which we will focus on in later chapters.

- **Wave 4+**: By the fourth wave, the automation program should be churning out automation projects like an assembly line. The development team should be working on the *must-do improvements* and the *long-term improvements*, but we should also be focusing on and prioritizing any incoming *quick wins* or *low-hanging fruit* that come organically. Like in wave two, as business units see the value of automation, it's not unheard of for those groups to start sending requests and ideas for new automation opportunities. However, these opportunities must still be assessed for best fit, feasibility, and viability before being placed into the backlog of ready-to-implement opportunities.

Even with our use case backlog prioritized and ready for implementation, we should always be actively looking for additional automation opportunities to keep the pipeline flowing – either by reviewing the results of our top-down approach and targeting new business units to showcase automation or casting a net by trying out more bottom-up approaches.

Looking at the end-to-end process

Now that we have our pipeline created and an implementation plan regarding how to prioritize the pipeline, we can get started with the first steps of implementation – reviewing the current state.

Because of the general low complexity of implementing RPA, it is common with many organizations to leverage automation within the current state of an existing process. In general, RPA does not require much, if any, process reengineering since RPA interacts at the desktop level. This makes it very easy to fall into the trap of automating the *as-is* process. Automating the *as-is* nature of a process ensures quick development efforts and low change management, which is something that new automation programs gravitate toward in order to showcase quick value, but we run the risk of missing out on any additional value and impact that could be generated if the *as-is* process was investigated and reengineered into a *to-be* process. Every project that involves RPA, whether it be a *quick win* or a *long-term improvement* from the previous section, should be combined with some level of process understanding, analysis, and reengineering. Therefore, it is important to review the current state of an automation opportunity with a process reengineering hat on; we do this by looking to optimize the process before applying automation and by looking upstream and downstream within a process.

Implementing automation in a particular process opens the window for revisiting the process as a whole and how it's executed. Often, existing business processes are overly complex, usually a result of an old process that was never fully examined, and full of unnecessary steps that could be eliminated. To add to the complexity, most business processes are not fully standardized, nowhere near optimal, and most often, not documented. These are all great reasons to apply process reengineering exercises. Reengineering the process can provide us with benefits, such as the following:

- **Process clarity**: Reengineering the steps of the process provides the opportunity to create updated process documentation – ensuring all colleagues understand the process and perform the process consistently.

- **Standardized procedure**: The removal of unnecessary process steps and the standardization of inputs and process steps can streamline operations and improve the efficiency of the business process.

- **Maximized impact**: By optimizing the process, we can more likely surpass the targeted business objectives of the business unit.

- **Maximized ROI**: By reviewing the big picture, we can potentially generate an additional, indirect, level of impact and value provided to process the upstream and downstream of the automation opportunity.

As part of understanding the automation opportunity, you should ask the process owner and SMEs for any existing documentation or standard operating procedures. Unfortunately, it is not uncommon for process knowledge and understanding to be quite low; what you might quickly start to realize is that a lot of these processes are not documented at all, and a lot of these processes are performed differently by each employee completing the task. Even with process documentation, you should schedule meetings with the SMEs to walk through the process live. We'll discuss, in more detail, how to run these process walk-throughs in later chapters, but it's important to hold walk-throughs with multiple SMEs completing multiple cases or variations within the process. Typically, each employee follows what they understand to be the best practice; this can lead to differing procedures by each employee. The benefit of capturing the steps of multiple employees is that you now get a glimpse of what steps work and what steps don't work within a process. This should ultimately lead to significant improvements in the process being automated, as we only consider the most optimal steps within a business process.

As part of current state process walk-throughs with SMEs, it's important to question decision points within the process or any judgment calls made by the SMEs. Most decision points within the process should be definable with a set of business rules; always question whether an SME says the rule for a decision point is based on experience – often, you can create a set of rules for an automation to follow. Unfortunately, there might be times when a judgment call is required and automation cannot be followed by a set of rules. In that case, we can potentially turn to the cognitive techniques of later chapters to alleviate that gap.

These process walk-throughs act as the first step in optimizing the process and getting it ready for automation. Conducting process walk-throughs will give us intimate knowledge about the process at a keystroke level, immersing us into the intimate details of the current state. It might be easier to be fully tunneled into the process and forget about the big picture; however, to truly optimize a process, and generate the most impact, it's important to consider what the process connects from and what the process connects to, that is, upstream processes and downstream processes. Taking a step back to look at the big picture and analyze the upstream dependencies and downstream impacts of a process opens up the possibility of identifying additional benefits and opportunities for greater value extraction.

Going back to our HR onboarding example from the *Identifying target goals* section, we can see that there are many subprocesses within the larger onboarding process (*Figure 4.11*). Most of these subprocesses likely have opportunities to integrate automation, but by focusing too much on one specific subprocess and not the entire process, we might not uncover the mammoth-sized opportunities that could generate a greater impact:

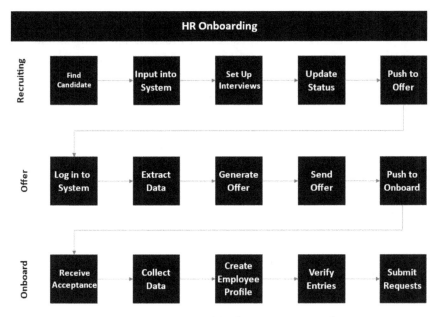

Figure 4.11 – A HR onboarding process example

When assessing upstream dependencies, we want to understand what bottlenecks or barriers exist that limit an initiative's value, such as the following:

- Are there opportunities for automation to make data cleaner/more standardized or to make data available more quickly?

- Are there non-value-add activities that can be eliminated?

- Are there policies that can be altered or removed?

Inversely, when assessing downstream impacts, we want to investigate whether there are any additional benefits of automating the opportunity outside of the targeted business objectives set by the business unit, such as the following:

- Do increased accuracy and data quality remove the need for QA-type roles?

- Can this increase in accuracy reduce regulatory fines and/or reduce time spent resolving errors?

- Are there additional design considerations that could optimize downstream work?

Taking a step back and viewing the process at a big-picture level allows us to view other processes beyond their pain points. Once we start to view RPA as more than just supplementing the *as-is* process, we can start to use RPA as a means of business improvement. By having that view on the end-to-end state of an automation opportunity, we unravel substantial savings that we could likely not touch by automating the *as-is* state alone.

To wrap up, our automation pipeline should be prioritized in a manner that quickly generates the enterprise value. In the early stages of a program, our focus should be on building out *quick wins* and *low-hanging fruit*, that is, opportunities lower in complexity that generate value. As we venture out and gain more experience, we can dive into the *long-term improvements* and the *must-do improvements*. To ensure that we deliver value, as we choose automation opportunities from the pipeline, it is important to first sit down with the SMEs and review the current state of the process. During this analysis of the current state, we should consider optimizing the process, as implementing RPA into a process is a great opportunity to add process improvement, if possible. As part of process improvement, we should always take a step back and investigate any upstream and downstream dependencies, as improving those dependencies can provide additional layers of impact that generate the enterprise value.

In this section, we reviewed how to evaluate potential automation opportunities with the best-fit, feasibility, and viability assessments. Then, we ventured into organizing the remaining opportunities that passed our assessments, prioritizing our pipeline. Once prioritized, we ventured into reviewing the full end-to-end state of the automation opportunity in order to grab a full understanding of the question at hand. These steps are all critical when trying to ensure our pipeline is filled with meaningful automations. As we venture into the next section, we will dive into the specific characteristics to probe for when assessing cognitive automation opportunities.

Probing for cognitive automation

When assessing automation opportunities, there likely will be opportunities that are too complex for traditional automation but are compelling for cognitive automation. In addition to the characteristics of cognitive opportunities discussed earlier (*probabilistic in nature*, *high variability*, or *unstructured data*), it is pivotal to ask the correct questions with the correct participants when assessing automation opportunities for cognitive automation.

For initial conversations about the automation opportunity, it's important to have the following participants available for the process workshops:

- Process owner
- Data scientist (if available)
- RPA solution architect
- Process SME
- RPA business analyst

During these process workshops, it is very important to have both the process owner and as many SMEs as possible available. Generally, the process owner has a lot of pull and say into how the process can be reengineered or improved, but typically, these process owners might not have too much day-to-day interaction with the process. The SMEs can help fill this void, as they can provide immediate feedback and considerations for improving the investigated process. It's important to create a good relationship with these SMEs, as they will likely be a very common contact when gathering requirements and implementing the opportunity.

As you build a relationship with these stakeholders, it's paramount to probe their availability during implementation. Many projects fail due to the lack of stakeholder availability, so ensure that the SMEs included in the process workshops have availability to assist during implementation. There are many pieces to creating cognitive automation, thus it is important to have SMEs that can help with the following:

- Current state analysis
- Future state solutioning
- Data gathering
- Requirements gathering
- Process documentation
- Documentation sign-off
- Data labeling
- Sprint demos/updates
- UAT testing and analysis
- UAT sign-offs
- Production monitoring

The preceding list is just some of the many areas where the SME will be needed to assist. Make sure they have enough time to dedicate to the effort.

The RPA solution architect and business analyst are very important to include in these workshops as they can further assess the feasibility of the opportunity and help work toward a future state design. The business analyst should be able to help probe into the data inputs that are necessary for the process, such as the following:

- **Document delivery**: Aim for use cases with consistent and accurate document delivery processes – ask about where the documents originate from and the data types. Strive for digital documents, if possible.

Manually scanned documents are feasible with automation and document understanding; however, first, try to check to see whether OCR can reliably extract the necessary information.

Inquire about the volume of documents delivered, as any variation in delivery can impact the complexity of automation. For example, if you have a use case where 80% of documents are saved to a shared drive, while the remaining 20% of documents are emailed to a shared inbox, then an RPA developer would have to create a connection to the shared drive to gather the majority of documents, in addition to filtering out an email inbox for a certain subject line or sender.

- **Document structure**: Focus on the structure of the incoming documents – ask about the potential of structural changes or whether a file can contain multiple documents. Strive for structured or semi-structured documents/data – unstructured documents/data can raise process complexity.

It's extremely important to ask about the structure of each type of input document. For example, if you have an invoice process, ask the client for as many different types of invoices from as many clients as possible. This variation of input data will allow you to review the similarities and differences of each template type. If it's not feasible to get every type of variation, ask for the invoice templates from the highest volumes to review.

Probe into process redesign if the current state contains inconsistent document structures. Never be afraid to consider process redesign. Often, it can greatly improve the success of a project, while at the same time, not being too difficult to complete.

- **Document extraction**: Probe what required fields need to be extracted from each document; often, fields can be deemed unnecessary for automation. We'll probe into this a little more when gathering the full requirements of the automation opportunity, but it's never too early to start working on the taxonomy of the automation process.

Focus on the data types of each field; question whether there are multiple languages or formats that need to be considered. Remember, handwriting can raise process complexity.

- **Document verification**: Ask for multiple copies of each type of document – small variations of documents can be unnoticeable to humans but can quickly raise process complexity.

Like document structure, primarily focus on the largest volume of document types for automation. Set expectations with the SMEs early on regarding data validation and human-in-the-loop capabilities.

Finally, a data scientist can be extremely beneficial to have on preliminary workshop calls. This is because they can help probe for what data considerations would be necessary to build out the opportunity, such as the following:

- **Data sample size**: Is there a large enough sample size to train a machine learning model?

  This is extremely important to know ahead of time. At times, gathering data can be very difficult for many reasons (for instance, insufficient access, security concerns, and corporate politics); therefore, it is important to know ahead of time that there is enough data available to work with. In later chapters, we'll discuss how much data to target, but remember, we'll need data to develop and test with, in addition to separate data to validate our development against.

- **Data noise**: Does the dataset contain irrelevant information or randomness?

  As you gather data, make sure you are careful of any noise or randomness. Noisy data can often produce undesirable results as our models can start to think of the noise as a pattern. This pattern can then lead to generalizations by the model and incorrect assumptions. If you start to see irregularities or randomness in the dataset, one of the easiest ways to resolve these issues is to inquire for more data.

- **Data completeness**: Does the data sample contain all the information necessary for modeling?

  In addition to data noise, it's important to ensure that the data samples available contain all the fields that are necessary for implementation. Sometimes, the data sample might only include a snippet of the number of fields necessary to complete the process, with the remaining data fields coming from other sources – make sure that you probe the SMEs about the relevant data fields necessary for the process and their locations in the sample data.

- **Data availability**: Do the stakeholders (process owners/SMEs) have access to produce data for modeling? Can production data be retrieved in a timely manner?

  Finally, like data sample size, it's important for us to be able to get data in a timely manner. For example, if our SME needs to request data from other sources, make sure that you have a sense of the average turnaround time – waiting for data to become available can have a large impact on the implementation timelines.

As we dive deeper into cognitive use cases, we will see the importance of these considerations, especially around source data. While you might not receive answers to some of these questions during early process workshops, keep these considerations in your back pocket as you jump into the implementation phases, as they will reappear when we discuss automation design, development, testing, and more in the upcoming chapters.

Summary

In this chapter, we discussed the beginning of any automation program: creating an automation pipeline. We learned about the characteristics to look for in potential automation and cognitive automation opportunities and how to lead discovery exercises to search for automation opportunities. Following this, we learned how to investigate whether an automation opportunity is the best option to implement, and then we learned how to dive deeper into the current state of the automation opportunity to optimize the process in order to capture the most value. Every automation program should have a similar goal of providing value to the business. By learning how to search for automation opportunities, how to understand the details of each opportunity, and what to probe for when dealing with cognitive automation, we can ensure the automation opportunities developed provide value to the end users that leverage them.

In *Chapter 5, Designing Automation with End User Considerations*, we will take the next step with our automation opportunities captured using the techniques of this chapter. We will learn how to gather user goals and requirements, and we will learn how to incorporate these goals into the design of a future state automation while properly scoping the process to reduce complexity.

QnA

1. What are the characteristics of RPA?

    - It requires manual interaction with applications, is repetitive in nature, has standard inputs, is time-consuming, is rules-based, and is structured.

2. What are the characteristics of an automation opportunity that requires cognitive technologies?

    - Probability in nature, high variability, and unstructured data

3. What exercises are performed when evaluating potential automation opportunities?

    - A best-fit assessment, a feasibility assessment, and a viability assessment

4. When building an automation, what wave would we place the following automation into?

    - An invoice processing automation that interacts with 2 applications, 7 screens, 20 process steps, no business variations, and does not require image-based automation.

    - Process complexity is low based on the characteristics provided; the opportunity can be placed in either wave one or wave two of the pipeline, depending on the potential value provided by the opportunity.

5. Can we proceed with a cognitive automation opportunity that contains a small sample size of data?

    - Yes; however, we need to be cautious. One of the selling points of leveraging UiPath's AI Center is the ability to close the feedback loop and continuously train cognitive technologies. Therefore, if the opportunity provides a large amount of value, but there is not much training data available, we can proceed with the opportunity with the expectation that the automation will learn and grow with human validation.

5
Designing Automation with End User Considerations

Once **automation** opportunities have been identified, the next milestone in the automation journey is focused on requirements gathering. During this phase, it is imperative to gather as much context as possible around the end user's requirements and goals for automation to ensure that **cognitive automation** can achieve the end user's **target goals**. Understanding the target goals of the use case will determine the future success of applying cognitive automation to an opportunity.

Every automation should be designed with the end user's considerations in mind. Automation that is difficult for the end user to interact with is automation that runs the risk of being unused; thus, in this chapter, we will first focus on gathering the end user's requirements, then move toward setting the target goals of future state automation. Finally, we will dive into how to design a future state automation solution.

In this chapter, we're going to cover the following main topics:

- Gathering requirements
- Setting target goals
- Designing the solution

Gathering requirements

The foundation of any successful automation is an accurate understanding of the *current state process* and an accurate understanding of the *requirements of the future state process*. Being in sync with the current state of the end-to-end process allows us to fully understand the intricacies of the automation opportunity. By gathering the requirements of the current state process, and understanding the broader objective of using automation, we can gather valuable insights to help shape the future state design of the opportunity.

Gathering the current state

During the requirements gathering phase of the automation life cycle, we start our journey by gathering the current manual steps an end user performs to complete the process we plan to automate. By gathering the details of how an end user manually completes the current state, we can search for inefficiencies within the process and start to evolve the process into a future state design, with automation and cognitive technologies incorporated.

In order to gather the steps of the current state, we need to work with individuals closest to the process. In the last chapter, we discussed how a process owner could help with sizing an opportunity for automation, and how they may not be the best resource to provide step-by-step instructions for how to complete the process task; thus, we likely need to interact with other individuals who are familiar with completing the task manually on a regular cadence, such as **subject matter experts** (**SMEs**). These SMEs will act as our main point of contact when gathering the current state, providing feedback on the future state design, and testing the effectiveness of the automation during **User Acceptance Testing** (**UAT**).

Figure 5.1 – Requirements gathering

Gathering the current state of an automation opportunity only takes a few steps:

1. Gathering current state documentation

2. Shadowing the process

3. Understanding the objective

Once these steps are completed, we should have enough context around how the current process is performed, and can then understand more accurately the objective of applying automation and a more effective solution for a future state design.

Gathering current state documentation

Most automation opportunities are to do with processes that humans manually complete as part of their day-to-day business activities. Because humans complete these tasks, documentation is created to capture how to perform the task, and it is usually leveraged to train new employees, or for auditing purposes. Other benefits of documenting business processes include the following:

- Auditability

- Business continuity

- Consistency and productivity among team members

As you probe for documentation, note that teams may refer to these documents by different types of names, such as the following:

- Desktop procedures

- **Standard operating procedures (SOPs)**

Not every team you encounter will have their tasks documented, so when gathering the current state, it's best to ask the *process owner* or the *SME* for any documentation they have on the process.

If no documentation on the process exists, this could be an opportunity for the SME to document the manual steps of the process, potentially using an application such as **UiPath Task Capture**. Task Capture allows the end user to create process documentation by either building a process flow diagram and populating it with screenshots, or by recording the manual steps completed by the end user, as shown in *Figure 5.2*. Once created, this current state documentation can then be leveraged as part of the automation's documentation.

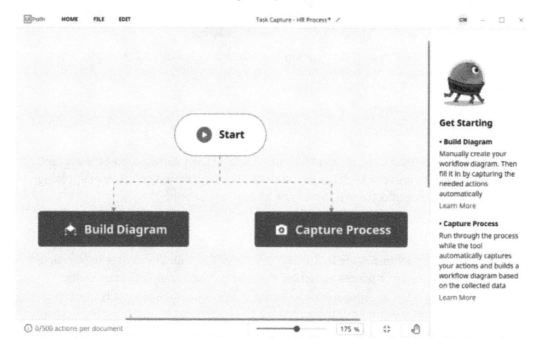

Figure 5.2 – Using UiPath Task Capture to capture the current state of a process

While gathering existing documentation can provide context and understanding into the manual steps of a process, it should not replace process shadowing; observing how an SME completes the process can be extremely impactful in how the future state of a process is designed, highlighting steps of the process that may be missed from any existing documentation. We will learn about it in the following section.

Process shadowing

One of the best ways to understand a process is to simply watch how the process is completed. By sitting with the SME, we can gather the details of the process through observation, while asking questions as they complete the process.

During **process shadowing**, we should always sit with the SME (either virtually or in person), observing the steps taken to complete the task at hand. During this observation, it is always recommended to record the session through a video conference, if possible. This allows us to have a reference to all the questions asked, or details of each step, without needing to take overly detailed notes. It also can be helpful when moving onto the development phase, as a recording of the manual process could augment any process documentation created for the automation (such as a **Process Design Document**).

Process shadowing gives us the luxury to probe the SME. We can ask questions such as the following:

- What works well?
- What causes difficulty?
- Why do we perform the task this way?

During observation, we should probe the SME for any *pain points* or *potential improvements* that could be included in the automation design. Many business analysts fall into the trap of capturing the current state of the automation without challenging any of the steps or decisions made during the process. Actively listen to SMEs for pain points and requests for the future state. Never make assumptions about how you think the end user interacts with a process, or what the end user wants in a future state. Instead, ask open-ended questions that allow the SME to completely provide their input. If the process is inefficient, challenge the SME on why the process is performed that way. These identified pain points should be addressed in the future state design of the automation.

Understanding the objective

As we uncover the current state details of an opportunity through process shadowing and documentation, we should be able to start understanding the broader objective of the process. By understanding the objective of the process, and understanding the end user's requirements, we are better equipped to fit automation when finding a solution for the future state. This allows us to deliver better and more impactful automation to the business.

Automation must be designed with the end user's experience in mind. As we shadow SMEs and gain an understanding of the current state, we should be keeping an eye on how the end user interacts with the current state process. By confirming the objectives and goals of the end user, we can gather a mutual understanding of what the project should achieve. Some questions that should be answered when understanding the objectives are as follows:

- What is the purpose of the task/process?

- What is automation set to achieve?

- What are the expectations of automation?

- How is the output of this task/process used by others?

As we understand the objectives of the automation project, it's important that we have the full picture of the end-to-end process prior to beginning development. As discussed in the last chapter, it's advantageous to have the full context regarding the upstream and downstream processes of an automation opportunity. By knowing how the process connects to other tasks/processes upstream and downstream, we can more accurately judge how cognitive automation can fit into the use case, potentially allowing us to alter the upstream or downstream processes to better fit with the automation.

By understanding the current state of the automation opportunity through current state documentation and process shadowing, and understanding the objective/needs of the end user, we can fully understand the opportunity for applying automation. With this newly acquired understanding, we can then continue with setting the right expectations and setting realistic goals for automation.

Setting target goals

Very often, the cause of unsuccessful projects is *unclear requirements*, leading to *unrealistic target goals*. With automation, the end user (and their experience) should always be the top priority when designing a project. Automation that is developed without achievable target goals is likely to be abandoned and turned into *shelfware*. To mitigate this abandonment, we must understand the end user's objectives for the automation, and then set realistic target goals for success.

Setting realistic target goals early in the development process is a large preventer of abandoned automation. Too often, automation deployed into production is abandoned by end users because it doesn't perform to expectations, or doesn't fully meet the end user's requirements or objectives (*Figure 5.3*):

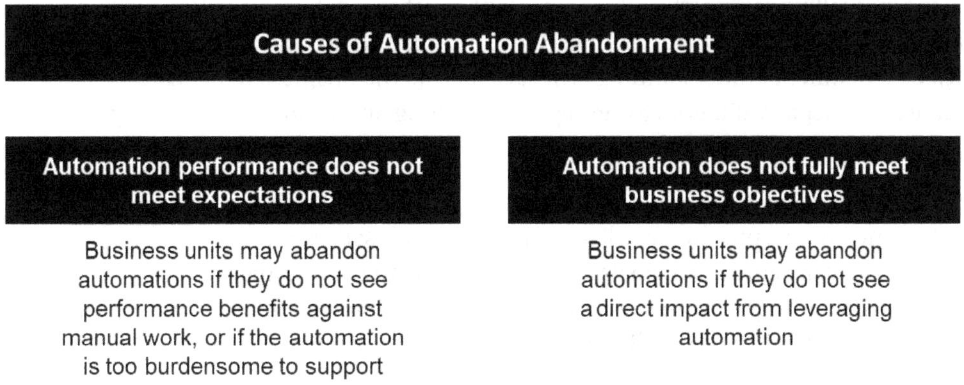

Causes of Automation Abandonment

Automation performance does not meet expectations	Automation does not fully meet business objectives
Business units may abandon automations if they do not see performance benefits against manual work, or if the automation is too burdensome to support	Business units may abandon automations if they do not see a direct impact from leveraging automation

Figure 5.3 – Causes of automation abandonment

Prior to starting development, align with the end users on a set of success criteria for the *to-be* (or future state) cognitive automation. Some examples of success criteria include the following:

- Time savings
- Efficiency gains
- Straight-through processing—the percentage of items (such as documents) that require no human intervention
- Field accuracy
- Ease of use
- Speed/throughput
- Total cost of ownership
- Vendor relationship

With objectives and goals for the project aligned and an understanding of the full end-to-end process completed, we can set realistic target goals for the automation opportunity. It's important to educate the end users about the success rate of cognitive technologies. Cognitive automation is not going to return a 100% accuracy rate. There is a chance for small variances, thus it is important to set the right expectations with end users early about the success rate of machine learning, **optical character recognition (OCR)**, or Document Understanding. This also allows the business to allocate human users to act as validators for exception handling and **human-in-the-loop (HITL)** capabilities.

By fully understanding the objective of the automation opportunity, gathering context on the full end-to-end process, and listening of the end user's requirements for automation, we can set realistic goals and expectations to the automation, and ensure proper documentation of the opportunity with UiPath templates, such as the **Process Definition Document (PDD)**. All these points help ensure a well-received product once deployed in production.

Designing the solution

Building an automation takes a lot of investment from a time and material perspective, so we must ensure that any automation deployed in a production environment is designed in a manner that incentivizes the end users of the business to continue using the automation. Focusing on the end user when designing the solution includes the following:

- Choosing the correct automation type
- Designing for the best user experience

By building an automation that's easy for the end user to interact with and understand, we can achieve our forecasted benefits, as the end user has a positive experience when interacting with the automation.

Choosing the correct automation type

When considering the design of a **cognitive automation**, one of the first decisions encountered is what automation type to leverage. With **robotic process automation**, there are two types of automation that impact the way end users interact with automation: attended and unattended automation (*Figure 5.4*), each providing different types of engagement with automation.

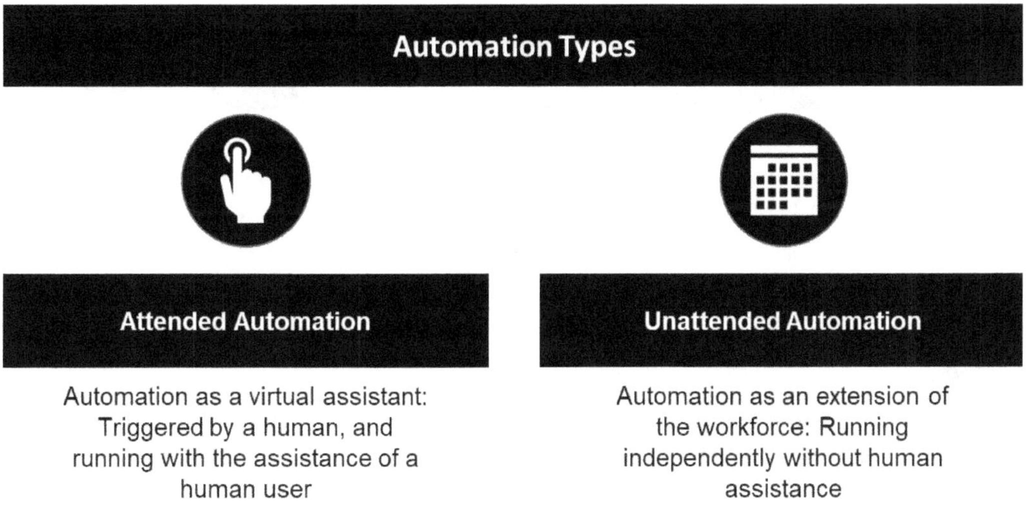

Automation Types

Attended Automation

Automation as a virtual assistant: Triggered by a human, and running with the assistance of a human user

Unattended Automation

Automation as an extension of the workforce: Running independently without human assistance

Figure 5.4 – Automation types

Both types of automation come with different sets of requirements and considerations for implementation. However, by acquiring a deep understanding of the end user's target goals, the choice of automation type should not be a difficult one to make.

Attended automation

Attended automation can best be seen as a virtual assistant to the human user. These types of automation are run directly from the end user's computer and must be triggered directly by the end user themselves. Attended automation tends to give the end user more control, as they get to choose when the automation runs and, at times, what data the automation can leverage. Additional points on attended automation are included in *Figure 5.5*:

Attended Automation *Virtual assistants, running with the assistance of a human user*	
What to Run	• Runs smaller, task-based jobs that are more "front office"
When to Run	• Can run on a regular basis, but are also "ad hoc" types of tasks, running to assist a human whenever needed
How to Run	• A human chooses to run the automation by pressing "play" in their UiPath Assistant
Where to Run	• Runs directly on the user's desktop, always available for the user to execute
Why to Run	• Best for shorter, more personal tasks that fulfill an individual's needs • Typically require human input (in other words to enter a file path, a date range, etc.) • Typically save smaller amounts of time as tasks are shorter (attended automations typically take only a few minutes to run)
When to Choose	• When developing for a problem that is variable for each user and not standard, this requires human input • Automation only provides smaller levels of benefits, usually not recovered by a full implementation effort

Figure 5.5 – Attended automation

Because attended automation is triggered by the human end user, it gives automation developers an excellent opportunity to juggle information between the robot and the human. This allows developers to add message prompts to their automations, allowing the automation to ask human users for input during runtime, such as the following:

- Prompting the human to input application credentials

- Prompting the human to choose a local file for input

- Prompting the human to input data that affects the logic of the bot (for example, the account number, and start and end dates)

These types of prompts also allow automations to prompt for human validation in case of exception. Through the use of UiPath's Validation Station, automation can ask the human to provide input when it faces a scenario of low confidence, such as not being able to categorize an email (*Figure 5.6*):

Figure 5.6 – UiPath Validation Station

With UiPath's Validation Station, we can provide a pop-up window on the user's desktop, asking them to provide input, or validation, on a particular item in a document. With these types of prompts, we can ask the end user for real-time input, placing automation as a real companion to the end user.

Considerations for using attended automation

With the ad hoc nature of attended automation comes a few considerations around documentation and performance. Since attended automation can potentially be shared with the entire workforce, proper documentation should be created to teach end users how to interact with the automation and the expected inputs and outputs of the automation. Providing sufficient context and information on an automation is imperative, making the automation easy to use, and enhancing the end user's experience.

To educate the end user on the context of attended automation, development teams can provide detailed project descriptions, user guides, and prompts within the automation to guide end users through execution. End users are more likely to consume automations that contain detailed descriptions of what the automation entails; thus, project descriptions should outline what the automation will perform, dependencies for running the automation, and the expected output (*Figure 5.7*). Teams should host these project descriptions on a common site, such as **UiPath Automation Hub** or **UiPath Marketplace**.

Figure 5.7 – Project descriptions, like those on Automation Hub, are best practice

In addition to documentation, leveraging UiPath prompts, such as callouts, can help guide the user through confusing, or difficult steps of the automation process (*Figure 5.8*). Callouts are extremely helpful as they can be placed next to elements on a user's desktop to *direct* the end user to interact with automation.

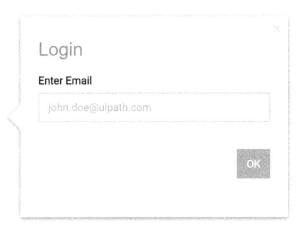

Figure 5.8 – Example of a callout in UiPath

In addition to the consideration around documentation, computing **performance** must also be considered when building an attended automation. Attended automations are best for short tasks, providing immediate output to the end user. These tasks should not take too much time, as attended automation may take over a user's machine while executing, leaving the human *computerless*. Leaving the end user computerless can be seen as a risk to potential benefit; if the end user can't perform other tasks while an automation is running, they may just wait for the automation to complete, reducing time savings. While there are technologies that allow for background processing (such as UiPath's **Picture-in-Picture** and **Simulate/Send Windows Messages Interaction Frameworks**), be sure to scope the attended automation precisely, keeping the impact on the user's machine minimal.

An attended automation is a great resource to leverage when dealing with short tasks that don't have a definitive scheduled runtime. Allowing the end user to trigger the automation gives the automation the chance to also prompt the user for input. However, with these features also come considerations around documentation and performance that can impact the end user's experience if handled incorrectly. With a lack of documentation, end users may not be convinced to try out the automation or may be confused on how to use it, while long-running attended automations may leave end users frustrated if they cannot use their computers for other tasks while the automation executes.

Unattended automation

Unattended automation can best be seen as an *extension of the human user*. This type of automation is run from dedicated machines (such as virtual machines on the cloud) and must be triggered directly by **UiPath Orchestrator** through a triggering mechanism, such as a schedule. Unattended automation tends to unlock higher levels of impact than attended automations, as placing an automation on a dedicated machine allows businesses to run automation(s) as often as needed. Additional points on unattended automation are included in *Figure 5.9*:

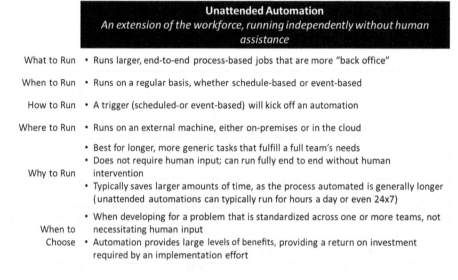

Figure 5.9 – An overview of unattended automation

Because unattended automation runs on external machines, we lose the ability to provide the end user with immediate prompts or guides, as we could with attended automation. On the bright side, UiPath has thought of this and has provided an alternative to Validation Station for unattended automations called **UiPath Action Center**.

Figure 5.10 – UiPath Action Center

With UiPath Action Center, an unattended automation can send **actions** to Action Center that can then be assigned to specific humans to review. This allows for similar prompting provided from Validation Station or from input dialogs, just with Action Center storing the action for the human to interact with. Once the human interacts with their assigned action, an unattended automation can then pick up the user's input and continue with its task.

Considerations for using unattended automation

With the scheduled nature of unattended automation come a few considerations around scheduling, documentation, and performance. Since unattended automations need a dedicated machine in order to perform automation jobs, there must be a machine available to run the automation, else the to-be performed automation must wait until a machine is freed up. If not planned correctly, this could lead to **service-level agreement** (**SLA**) issues with items not being completed in a timely manner. To resolve this, there could be a one-to-one mapping between automations and machines; that way, an automation always has a machine available when it's scheduled to run. Unfortunately, this could bring up issues with machine allocation and licensing, as additional machines and licenses will incur additional costs to the automation program. Automation programs should carefully balance automation schedules based on the machines they have provisioned.

In addition to scheduling challenges, documentation and performance are additional considerations for unattended automations. While the documentation for unattended automations doesn't need to tell the end user how the automation works, it should tell the end user what to expect from the automation and how to handle issues, should exceptions arise. Performance-wise, it can be easy to over-allocate an unattended automation with transaction items, only to see that the bot cannot perform as many items as originally planned. To alleviate this, you need to investigate leveraging queues, allowing multiple bots to perform the task simultaneously.

An unattended automation is a great resource to leverage when dealing with longer processes that can be scheduled. It allows the business to have an automation running in the background, and to complete transactions more timely, efficiently, and with greater frequency than a human user. However, with these features also come considerations around documentation and performance that can impact the end user's experience and increase the implementation cost if handled incorrectly. With a lack of documentation, end users may not understand how to handle potential exceptions or errors from automation, leading to support requests that can burden a support team. This lack of documentation can also lead to a misunderstanding about the types of exceptions raised by unattended automation, where specific exceptions must be handled by the end user. This misunderstanding can leave a bad impression, as end users may lose faith in the performance of the automation.

Designing automation for the best user experience

After choosing the type of automation that best fits the automation opportunity, we can move further into designing the overall solution. Successful automation implementations make it easy for users to engage and interact with automation; *the goal of automation is to augment human capacity, not burden it with additional rules and procedures to follow*. Thus, when designing for the best user experience, we should explore the inputs and outputs of the automation, and their interaction with the human user.

Providing data is a common ask for humans to perform prior to an automation beginning its execution. This type of input could be data from a system, such as a Case ID of an opportunity in Salesforce, or very commonly a path to a file saved to a shared or local drive. Many automations ask the human user for an exact input, such as the following:

- An exact filename (`ME_Report_MM.DD.YYYY.xlsx`)

- An exact wording convention (`Subject: Month End Report MM.DD.YYYY`)

While these exact inputs make it easier for the automation to pick up input data, it can dampen the experience for the end user, especially in the case of an uncommon naming convention leading to business exceptions; thus, when we need automation to prompt for human input, we should make it easy for end users to provide the necessary information. The following are some examples:

- Asking users to drop files into a network drive, irrespective of the filename. Automation should be able to deal with classifying the files during execution.

- Asking users to send emails with unstandardized subject lines. Automation should be able to deal with subject line headers and naming convention irregularities.

In the same way, the output of automation should also be easy for the end user to interact with. As mentioned earlier, the goal for automation is to augment human capacity, so we should aspire to have automation easily output its results back to its human counterparts, such as the following:

- Dropping its results into an organized folder structure

- Providing understandable output reports to interested parties

- Providing descriptive, yet understandable error messaging

- Providing steps for manually recovering failed items

- Providing easily accessible support in case of larger automation failures

Lastly, as mentioned in earlier sections, providing documentation on the automation is extremely important in uplifting the end user's experience. Documentation should be easily accessible and digestible, allowing the end user to provide *self-service* in the scenario that their automation performs against expectations.

In conclusion, the easiest way to ensure automation adoption is to design an automation that is easy for the eend user to interact with. An automation that's burdensome or frustrating to use will likely be abandoned in favor of the manual completion of the task. Designing the automation to easily allow the user to input data into the automation and interact with the automation's output is a recipe for successful automation.

Summary

In this chapter, we discussed the next phase of the automation journey: gathering requirements and designing the solution. We learned about how to gather requirements for an automation opportunity, how to set target goals for the automation, and how to design the automation with the end user experience in mind. By gathering the requirements of the business, reaffirming them by setting target goals, and incorporating this feedback into the future state design, we can ensure a successful automation deployment.

In *Chapter 6, Understanding Your Tools*, we will take the next step with our automation opportunities with the development phase of the automation life cycle. In the next chapter, we will learn about the characteristics of UiPath's cognitive tools, **Document Understanding**, **AI Center**, and **Computer Vision**, by venturing into their features and learning how to build with each tool.

6
Understanding Your Tools

Every great craftsperson is highly familiar with all the tools in their tool belt. An expert in the trade, the craftsperson knows exactly which tool is needed to get the job done. In our case, it is important for us to understand the tools available to us before developing a cognitive automation use case. In this chapter, *Understanding Your Tools*, we will learn about the characteristics of UiPath's cognitive tools:

- Document Understanding
- AI Center
- Computer Vision

By having a full understanding of the tools available to us, we can more accurately choose the right tool for the job. In this chapter, we'll learn about the Document Understanding activities and how they come together with the Document Understanding framework, which we'll use to develop Document Understanding use cases. We will then transition to AI Center, learning about the AI Center activities and touring the features of the tool. Lastly, we'll venture into Computer Vision, learning about the activities available to us with the tool.

In this chapter, we're going to cover the following main topics:

- Getting started with UiPath Document Understanding

- Getting started with UiPath AI Center

- Getting started with UiPath Computer Vision

Technical requirements

All code examples for this chapter can be found on GitHub at `https://github.com/PacktPublishing/Democratizing-Artificial-Intelligence-with-UiPath/tree/main/Chapter06`.

Working with UiPath is very easy, and with the Community version, we can get started for free. However, for UiPath AI Center, we will need an enterprise license. One can acquire a 60-day enterprise trial license from `uipath.com`.

With the 60-day enterprise trial, we will have access to the following:

- Five RPA Developer Pro licenses – named user licenses include access to Studio, StudioX, Attended Robot, Apps, Action Center, and Task Capture

- Five Unattended Robots, five Testing Robots, and two AI Robots

- AI Center, AI Computer Vision, Automation Hub, Data Service, Document Understanding, and Insights

For this section, you will require the following:

- UiPath Enterprise Cloud

- UiPath Studio 2021.4+

- The `UiPath.IntelligentOCR.Activites` package v.4.13.2 or higher

- The `UiPath.DocumentUnderstanding.ML.Activites` v1.7.0 or higher

> **Important Note**
>
> Directions on how to install UiPath packages can be found at `https://docs.uipath.com/studio/docs/managing-activities-packages`.

Enabling AI Center in the UiPath enterprise trial

UiPath Document Understanding comes out of the box with the UiPath Community and Enterprise versions; however, with UiPath AI Center, we need to enable the service within the Enterprise trial. You can enable the service by following these steps:

1. Navigate to **Automation Cloud**:

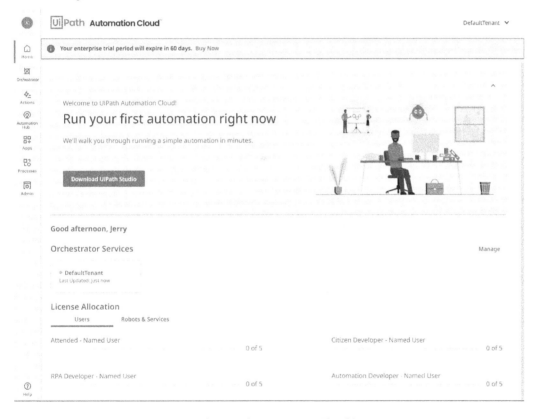

Figure 6.1 – The UiPath Automation Cloud home page

2. Navigate to **Admin**:

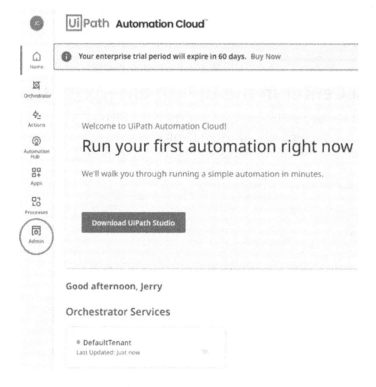

Figure 6.2 – Click on Admin

3. In **DefaultTenant**, click on the three dots, and then click **Tenant Settings**:

Figure 6.3 – Click on Tenant Settings

4. Choose **AI Center** under **Provision Services** and click on **Save**:

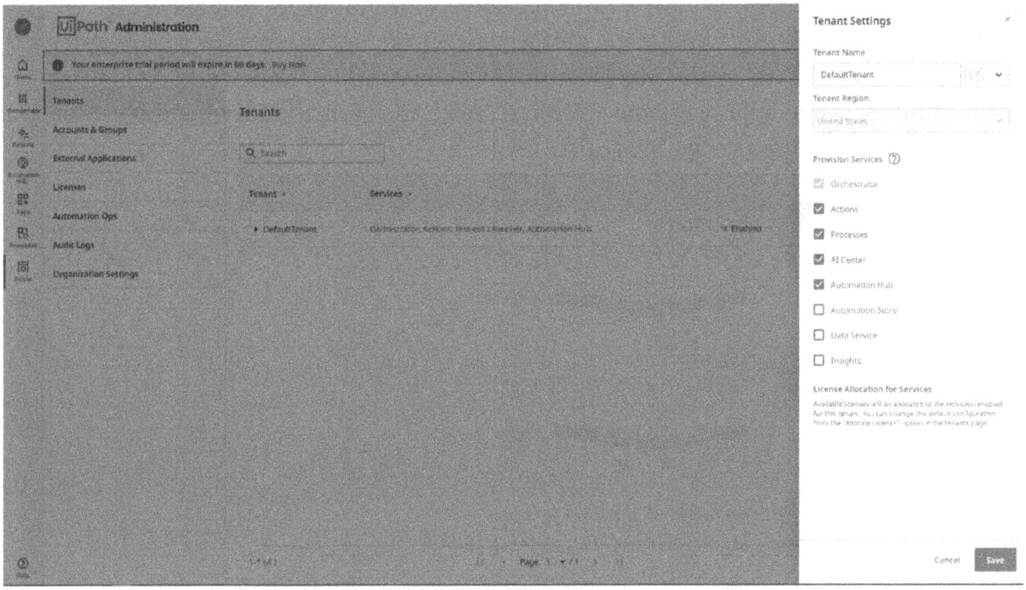

Figure 6.4 – Check AI Center and then click Save

Once you have AI Center enabled in **Automation Cloud**, we are ready to begin jumping into the cognitive tools of UiPath. The first tool we'll explore is UiPath Document Understanding.

Getting started with UiPath Document Understanding

UiPath Document Understanding unleashes the capability to intelligently process documents with Robotic Process Automation. With Document Understanding, we'll be able to digitize, classify, and extract data from PDFs, scanned documents, and even handwriting.

In this section, we'll learn about how to build Document Understanding use cases with the Document Understanding framework and the Document Understanding Studio template. We will lastly dive deeper into the activities or building blocks of the Document Understanding activity package.

Introducing the Document Understanding framework

Under the hood, Document Understanding is a series of technologies that allow for automation to read, interpret, and extract information from documents. To encapsulate these technologies, UiPath created the Document Understanding framework, combining all the technologies and approaches necessary to create an automation leveraging intelligent document processing (*Figure 6.5*):

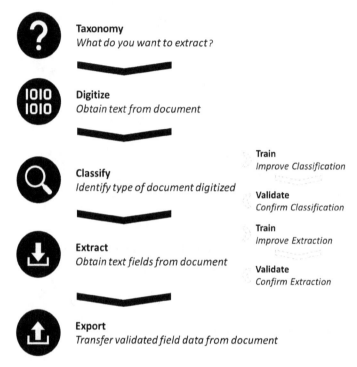

Figure 6.5 – UiPath's Document Understanding framework

Let's dive deeper into the Document Understanding framework and its five components:

- **Taxonomy**: The naming classification defined to classify the fields needed from target documents

- **Digitize**: Processing documents (with the use of OCR, if necessary) to extract all the text from target documents

- **Classify**: Classifying the document based on the text extracted from digitization

- **Extract**: Obtaining data by assigning text digitized to variables

- **Export**: Finalizing data into a format usable by automation

With the Document Understanding framework introduced, let's dive deeper into each component, starting with taxonomy.

Introducing the taxonomy

The first step in the Document Understanding framework is creating a taxonomy. A taxonomy is a system classification, and with Document Understanding, a taxonomy is a classification of **document types**. A document type is the classification of a logical type of document handled by a business process. Examples of document types can include the following:

- Invoices
- Tax forms (W-2 and 1099)
- Contracts
- Records

The Taxonomy Manager

During the creation of the taxonomy, we will categorize the document types needed and identify the fields within each document type – this will essentially become the schema (structure) for our data. To create our taxonomy, we will use the **Taxonomy Manager** within UiPath Studio to create a taxonomy of fields and document types. The Taxonomy Manager can be found in the Design ribbon of UiPath Studio (*Figure 6.6*):

Figure 6.6 – UiPath's Taxonomy Manager within the Design ribbon

---

Important Note

The Taxonomy Manager requires the `UiPath.IntelligentOCR. Activites` package installed as a dependency for your project. Once installed, a **Taxonomy Manager** button will appear in the Design ribbon of Studio.

Within **Taxonomy Manager**, we'll have the opportunity to define **Document Types**, **Document Type Details**, and data fields, as shown in *Figure 6.7*:

Figure 6.7 – UiPath's Taxonomy Manager

- **Document Types**: Define your document types, organized under your defined groupings.

- **Document Type Details**: Define the document type details and any number of fields within the document.

- **Edit Field**: Define the data type of the field.

Creating a taxonomy

Let's get started with creating a taxonomy! Start a new project in UiPath Studio and open `Main.xaml`.

> **Important Note**
>
> Before starting with this exercise, make sure you have the following packages installed: `UiPath.IntelligentOCR.Activites package v.4.13.2` or higher and `UiPath.DocumentUnderstanding.ML.Activites v1.7.0`.

1. Download the invoice example from the GitHub repository: `https://github.com/PacktPublishing/Democratizing-Artificial-Intelligence-with-UiPath/blob/main/Chapter06/Document_Understanding_Test/Invoice%20Template.pdf`:

Invoice

Invoice Number	INV-3337
Order Number	12345
Invoice Date	January 25, 2016
Due Date	January 31, 2016
Total Due	**$93.50**

From:
DEMO - Sliced Invoices
Suite 5A-1204
123 Somewhere Street
Your City AZ 12345
admin@slicedinvoices.com

To:
Test Business
123 Somewhere St
Melbourne, VIC 3000
test@test.com

Hrs/Qty	Service	Rate/Price	Adjust	Sub Total
1.00	Web Design This is a sample description...	$85.00	0.00%	$85.00

Sub Total	$85.00
Tax	$8.50
Total	$93.50

ANZ Bank
ACC # 1234 1234
BSB # 4321 432

Payment is due within 30 days from date of invoice. Late payment is subject to fees of 5% per month.
Thanks for choosing DEMO - Sliced Invoices | admin@slicedinvoices.com
Page 1/1

Figure 6.8 – An invoice example

2.   Open the Taxonomy Manager from the Design ribbon.

3.   Add a new group, **Forms**, by clicking + to the right of **Any Group**:

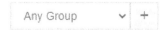

Figure 6.9 – Click + to add a group

4.   Add a new category, **Invoices**, by clicking + to the right of **Any Category**.

5. Add a new document type, **Sliced Invoices** (the name of the example invoice type), by clicking the **Add New Document Type** button:

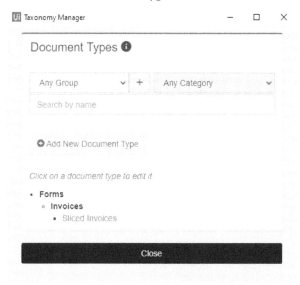

Figure 6.10 – The Taxonomy Manager after adding Forms | Invoices | Sliced Invoices

6. Create a data field by clicking on **New Field** under **Document Type Details**. Within **Edit Field**, add a name and a type of the corresponding data field. Click **Save** when completed:

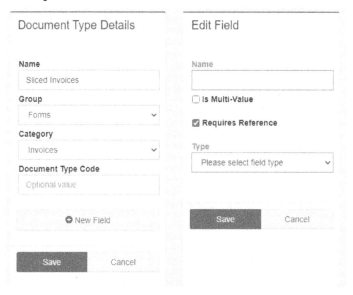

Figure 6.11 – Creating a new field

7. Fill the fields for the following data points (*Figure 6.12*):

- **Invoice Number** (text)

- **Order Number** (text)

- **Invoice Date** (date)

- **Due Date** (date)

- **Total Due** (number)

- The **From** address (text)

- The **To** address (text)

- The list of services

- **Sub Total** (number)

- **Tax** (number)

- **Total** (number)

Invoice

From:
DEMO - Sliced Invoices
Suite 5A-1204
123 Somewhere Street
Your City AZ 12345
admin@slicedinvoices.com

Invoice Number	INV-3337
Order Number	12345
Invoice Date	January 25, 2016
Due Date	January 31, 2016
Total Due	$93.50

To:
Test Business
123 Somewhere St
Melbourne, VIC 3000
test@test.com

Hrs/Qty	Service	Rate/Price	Adjust	Sub Total
1.00	Web Design This is a sample description...	$85.00	0.00%	$85.00

	Sub Total	$85.00
	Tax	$8.50
	Total	$93.50

ANZ Bank
ACC # 1234 1234
BSB # 4321 432

Payment is due within 30 days from date of invoice. Late payment is subject to fees of 5% per month.
Thanks for choosing DEMO - Sliced Invoices | admin@slicedinvoices.com
Page 1/1

Figure 6.12 – An invoice example

8. Once all the data fields are added, click **Save** and close the Taxonomy Manager – the
    results should mimic *Figure 6.13*:

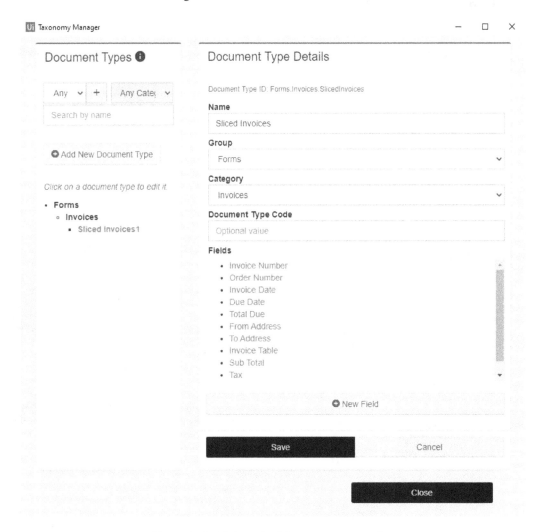

Figure 6.13 – The result of the taxonomy creation

9. Lastly, add a **Load Taxonomy** activity to `Main.xaml`, creating an output variable named `Taxonomy`:

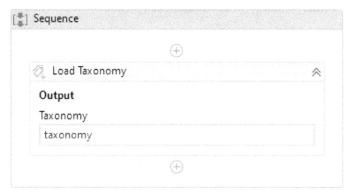

Figure 6.14 – Adding Load Taxonomy to Main.xaml

That's it! Our taxonomy is created and will be loaded into the automation at runtime. Note that in your project folder, a `Taxonomy.json` file has been created. This file can be saved and used in other projects to build a central taxonomy over time.

In the next section, we'll learn about digitizing the invoice document.

Introducing digitization

The next step of the Document Understanding process is digitization. Digitization is the process of taking input data and scanning it into machine-readable text. Once the data is digitized, we can classify and extract the relevant data fields to create actionable steps for automation.

Digitization in the Document Understanding framework has two outputs:

- The text from the input file (stored as a string variable)

- The **Document Object Model** (**DOM**) of the file – the DOM is a JSON object containing detailed information about the document

While digitization is the process of taking input data as machine-readable text, it is not just **Optical Character Recognition** (**OCR**). Digitization must be done with any type of input, digital and physical – any file that is already native text (unscanned PDFs or text files) can be read programmatically without the use of OCR.

For documents that require OCR (such as scanned PDFs), an OCR engine can be used during the digitization process. OCR engines that are compatible with digitization in the UiPath Document Understanding framework are as follows:

- UiPath Document OCR
- OmniPage OCR
- Google Cloud Vision OCR
- Microsoft Azure Computer Vision OCR
- Microsoft OCR
- Tesseract OCR
- ABBYY Document OCR

Let's continue with reviewing the Digitize Document activity before moving on to the exercise.

Digitize Document activity

For UiPath to digitize a document into a DOM, the **Digitize Document** activity (*Figure 6.15*) must be used within our automation project:

Figure 6.15 – The Digitize Document activity using the OmniPage OCR engine

Within the **Digitize Document** activity, notice the **Input** and **Output** parameters:

- **Input**:

  - **Document Path**: The full path of the document that the activity is set to digitize

- **Output**:

  - **Document Text**: Outputting the string text of the digitized document

  - **Document Object Model**: The corresponding DOM of the input file

In the next section, we will leverage the Digitize Document activity as we try digitizing a document.

Digitizing the document

Let's continue our example with digitization:

1. Drag a **Digitize Document** activity into the workflow after **Load Taxonomy**.

    Within the **Digitize Document** activity, we need to insert a file path as input and assign variables as output to hold the document text and the DOM.

2. Insert the path to the invoice document as the **Document Path** input and create a variable named document_text as the **Document Text** output. Lastly, add a variable named dom as the **Document Object Model** output:

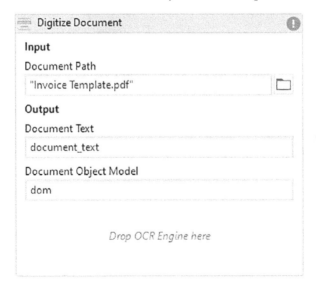

Figure 6.16 – Inserting the Digitize Document activity into our workflow

Note that we still have an error message, prompting us to insert an OCR engine. In this example, our document has the text embedded into the document (we can highlight and copy text directly from the PDF), so we don't need to rely on the OCR but still need to drag one into the activity to remove the error message. The OCR engine will only be used if the incoming document requires OCR processing.

3.   Drag **Tesseract OCR** within the **Drag OCR Engine here** region of the Digitize Document activity.

> **Important Note**
> More information on the OCR engines supported by UiPath can be found at `https://docs.uipath.com/document-understanding/docs/ocr-engines`.

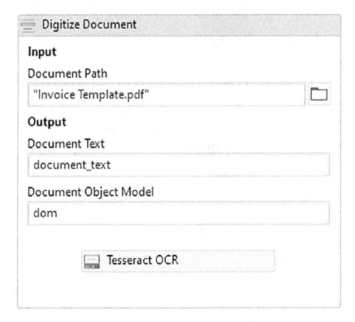

Figure 6.17 – Adding the Tesseract OCR engine

Within the **Digitize Document** activity, we'll keep the **ForceApplyOCR** flag as **False**, as we don't need to use OCR for this invoice because it's an embedded PDF. If we were to set the flag to **True**, then OCR would be used on the embedded PDF.

We will also leave the **DegreeOfParallelism** property to 1. By default, the property is set to 1. If we were to set it to -1, UiPath uses the number of cores on the machine to process pages in parallel. Any other positive value uses that specific number of logical processors to process pages.

Additional information on the Digitize Document activity and when to use OCR can be found at `https://docs.uipath.com/document-understanding/docs/digitization-overview`.

We have now loaded the taxonomy and digitized our invoice document. In the next section, we'll learn about classifying the invoice document.

Introducing classification

After digitizing our input data into machine-readable text, our automation can now interpret the text within our input document. However, while our automation can read the text, it still does not know what type of document it read; thus, the next step is to have the automation classify the text. Classification in the Document Understanding framework is using digitized input data to identify what type of document is input.

Document classification becomes extremely important when an input document contains multiple document types (*Figure 6.18*):

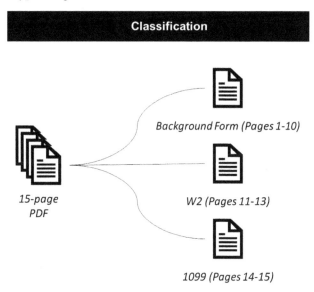

Figure 6.18 – Classifying documents from an input

For example, if we were to have a 15-page PDF containing a mixture of documents such as background forms, W2s, and 1099s, then with classification, we can split the 15-page document into small subsets based on the document type. If a file contains one document type, then classification will return just one classification result, but if a file contains multiple document types, then the classification result will return the document types found, along with their corresponding page ranges.

In the next section, let's look at the Classify Document Scope activity.

Introducing the Classifiers Scope

Classification of a document is performed via the **Classify Document Scope** activity:

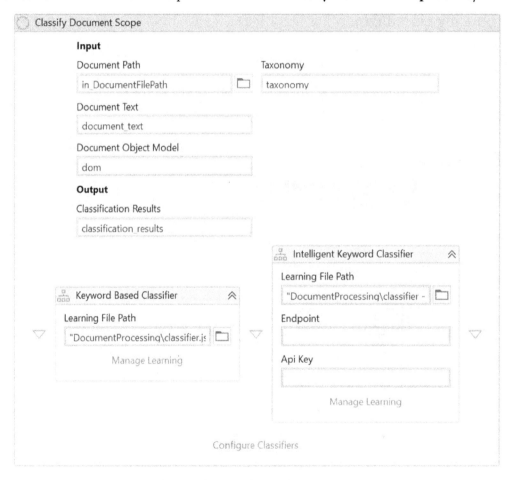

Figure 6.19 – The Classify Document Scope activity

Within the **Classify Document Scope** activity, notice the **Input** and **Output** parameters:

- **Input**:

  - **Document Path**: The full path to our target document

  - **Taxonomy**: The taxonomy returned from the Taxonomy Manager

  - **Document Text**: The document text returned from digitization

  - **Document Object Model**: The DOM returned from digitization

- **Output**:

  - **Classification Results**: The results stored in an
    `IReadOnlyList<ClassificationResult>` object, containing results such
    as classification confidence and OCR confidence (*Figure 6.20*):

```
Local Value                                 □   ×

ClassificationResult[1]
{
  ClassificationResult
  {
    ClassifierName="Keyword Based Classifier",
    Confidence=0.6877654,
    DocumentBounds=ResultsDocumentBounds
    {
      PageCount=1,
      StartPage=0,
      TextLength=848,
      TextStartIndex=0
    },
    DocumentId="Invoice Template.pdf",
    DocumentTypeId="Forms.Invoices.SlicedInvoices",
    OcrConfidence=1,
    Reference=ResultsContentReference
    {
      TextLength=15,
      TextStartIndex=77,
      Tokens=ResultsValueTokens[1]
      {
        ResultsValueTokens
        {
          Boxes=Box[4]
          {
            T:133.8024,L:76.7432,W:23.9624,H:14.1064,
            T:133.8024,L:103.152,W:32.2784,H:14.1064,
            T:133.8024,L:76.7432,W:23.9624,H:14.1064,
            T:133.8024,L:103.152,W:32.2784,H:14.1064
          },
          Page=0,
          PageHeight=841,
          PageWidth=595,
          TextLength=15,
          TextStartIndex=77
        }
      }
    }
  }
}

Copy to Clipboard                              Close
```

Figure 6.20 – An example of the classification results

With the Classification Scope activity introduced, let's continue with the Classifiers Wizard region within the Classification Scope activity.

Introducing the Classifiers Wizard

Within the Classifiers Scope, you will notice an area to insert different classifiers; this allows us to leverage multiple classifiers based on priority, as shown in *Figure 6.21*:

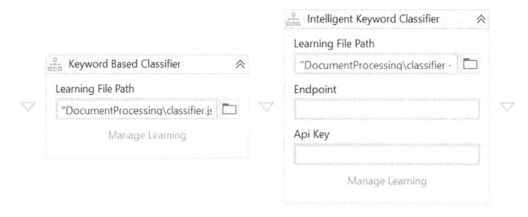

Figure 6.21 – Leveraging multiple classifiers at once

Within the Classification Scope activity, we place **Keyword Based Classifier** and **Intelligent Keyword Classifier** together to provide a more robust level of classification. We'll cover the different types of classifiers later, but let's configure a couple of classifiers with the Classifiers Wizard.

By clicking the **Configure Classifiers** button in the Classifiers Scope, we are greeted with a wizard that showcases our two classifiers, and our document types defined in our taxonomy (*Figure 6.22*):

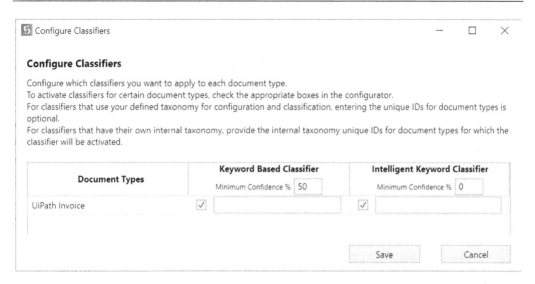

Figure 6.22 – The Classifiers Wizard

In this example, I have only defined one document type in the taxonomy (**UiPath Invoice**); hence, only one row is shown in the wizard. To the right of the **Document Types** column, note our two classifiers. Underneath each classifier, we have a minimum confidence percentage and a checkbox:

- **Minimum Confidence %**: This is a threshold of acceptable confidence from our classifiers. In the last section, as part of the classification result, we mentioned confidence as an output. With the minimum confidence %, we can control reporting based on the confidence return.

  In *Figure 6.22*, we set the **Minimum Confidence %** value of **Keyword Based Classifier** as 50. This means that if the Keyword Based Classifier was able to classify our document with a value greater than or equal to 50% confidence, then it will output that classification. If confidence was less than 50%, the Classify Document Scope will ignore that classification result and not report on it.

  In this example, if confidence was less than 50%, the Keyword Based Classifier's result would not be reported. The Intelligent Keyword Classifier's result would then be reported because it has a confidence threshold of 0%.

> **Important Note**
> Classifiers are executed with priority; those on the left are the highest priority.

- **Checkbox**: The checkbox underneath each classifier simply toggles whether you want to use a classifier for a specific document type. In the following example (*Figure 6.23*), note that **Keyword Based Classifier** is enabled for documents of the **Type1** document type, while **Intelligent Keyword Classifier** is enabled for documents of the **Type2** document type:

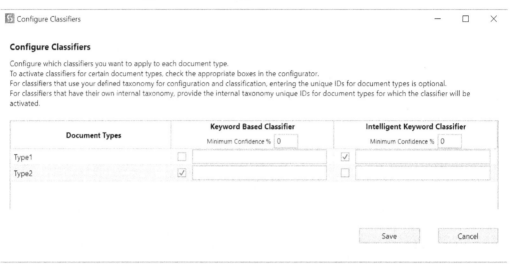

Figure 6.23 – Enabling classifiers for different document types

In this section, we reviewed the Classifiers Wizard with the Keyword Based Classifier and the Intelligent Keyword Classifier. In the next section, we will review the different types of classifiers offered in Document Understanding.

Reviewing the types of classifiers

Based on the requirements of your use case, you may be drawn to use one or more of the following classifiers available in Document Understanding:

- The Keyword Based Classifier
- The Intelligent Keyword Classifier
- The FlexiCapture Classifier
- The Machine Learning Classifier

The **Keyword Based Classifier** is a simple classifier that searches for repetitive keywords within a document. The classifier can work with one or more sets of keywords to match keywords. This classifier is best for classifying documents of a singular document type (no file splitting) and is best if the target document contains matching keywords within the first three pages of the file.

The **Intelligent Keyword Classifier** leverages an algorithm to intelligently classify documents based on the word vector it learns from document types. The classifier works by finding the closest matching word vector and reporting on the highest scoring document type match. This classifier is best for classifying documents of multiple document types (file splitting) and if the target document types are relatively easy to distinguish from each other.

The **FlexiCapture Classifier** is built by Abbyy to classify documents based on Abbyy FlexiCapture definitions. This classifier can be found within the `UiPath.Abbyy.Activities` package.

The **Machine Learning Classifier** leverages an ML skill deployed to UiPath AI Center to classify documents.

While these classifiers are very powerful and useful, there are times when human intervention is required to validate the output of the classifier. In the next section, we will discuss how to validate classification results with Document Understanding.

Validating classifiers

Sometimes after classification, we may require human validation of classifier output with the use of the **Classification Station**. The use of the Classification Station is generally recommended when dealing with multiple document types within a single file, or when a use case requires 100% accuracy in classification – it is always best to have a human perform a quick check before returning to automation.

The Classification Station is available for both attended and unattended automation through the use of the *Present Classification Station* activity for attended, and the *Create Document Classification Action* and *Wait for Document Classification Action and Resume* activities of UiPath Action Center.

When presented with the Classification Station, the user will have the ability to review and correct classifications performed by automation. In the Classification Station, the user can rearrange pages by clicking and dragging pages, and validate classifications performed by classifiers (*Figure 6.24*):

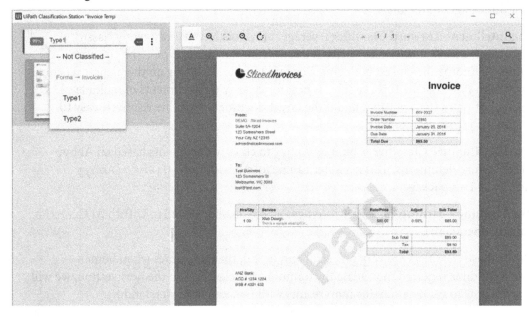

Figure 6.24 – Validating the classified invoice using the Classification Station

> **Important Note**
> More information on the Classification Station can be found in the UiPath docs: https://docs.uipath.com/activities/docs/present-classification-station#using-classification-station.

With our classification validation completed with the Classification Station, we can leverage our human validation to retrain and strengthen the classifiers used. In the next section, we'll discuss how to train classifiers.

Training classifiers

One of the great features of the Document Understanding framework is the ability to close the feedback loop and retrain classifiers after validation. Leveraging human input is a great way for automation to teach itself and improve performance for future executions. The training of classifiers is performed through the **Train Classifiers Scope** activity, as shown in the following screenshot:

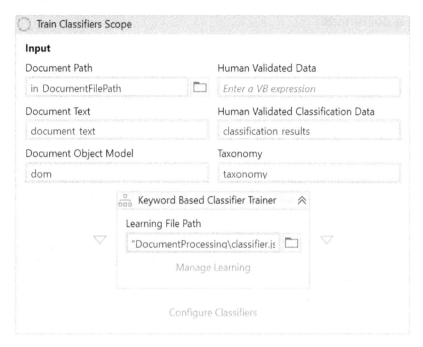

Figure 6.25 – Train Classifiers Scope

Within the **Train Classifiers Scope** activity, notice the **Input** parameters:

- **Document Path**: The full path to our target document
- **Document Text**: The document text returned from digitization
- **Document Object Model**: The DOM returned from digitization
- **Human Validated Classification Data**: The Classification Station results
- **Taxonomy**: The taxonomy returned from the Taxonomy Manager

Within the Train Classifiers Scope activity, you can train multiple classifiers (such as the Classification Scope), using the **Configure Classifiers** option to customize which document types are sent to training by a certain classifier trainer.

The following classifiers are retrainable:

- Keyword Based Classifier
- Intelligent Keyword Classifier
- Machine Learning Classifier

Classifying a document

Let's continue our example with classification.

In this example, we only have one invoice type on a page, so classification is not completely necessary. However, if our input file contains multiple documents (such as multiple invoices), then classification is required:

1. Drag a **Classify Document Scope** activity into the workflow after **Digitize Document**.

    Within the **Classify Document Scope** activity, we need to insert our file path, taxonomy, document text, and DOM as input, and assign a variable as output to hold the classification result.

2. Insert the path to the invoice document as the **Document Path** input, and insert our taxonomy, document text, and DOM variables into the **Taxonomy**, **Document Text**, and **Document Object Model** inputs.

3. Create a variable called `classification_results` and insert it into the **Classification Results** output:

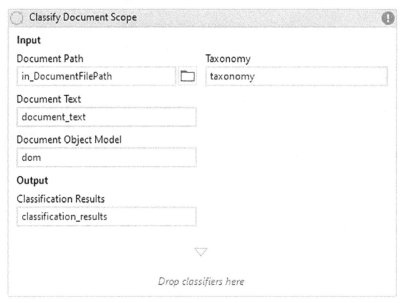

Figure 6.26 – The Classify Document Scope activity before adding a classifier

Since this document is an embedded PDF, we can use the Keyword Based Classifier for classification. The `Sliced Invoices` template appears very standardized, with the text `Sliced Invoices` appearing at the beginning of the document, so we can use that as our keyword to match.

4. Save a blank JSON file called `classification.json` in the project folder. This JSON file will be used as we manage the learning of the Keyword Based Classifier.

5. Click **Manage Learning** and choose our recently created `classification.json` file to start learning.

    Within the Keyword Based Classifier learning window, note the document type (`Sliced Invoices`) from our taxonomy. To start the learning process, we need to include common keywords within the document type. As mentioned earlier, the text `Sliced Invoices` is consistent with this type of invoice, so let's use that as a keyword.

6. Click **Add new keyword set**, and add the keywords `"slicedinvoices"` and `"sliced invoices"` (*Figure 6.27*):

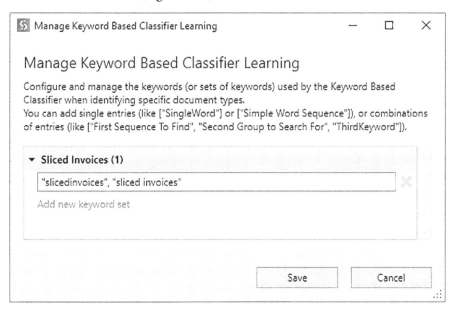

Figure 6.27 – Learning the Keyword Based Classifier

Let's next configure our classifier. Configuring our classifiers will be extremely important if we leverage multiple classifiers at once; however, in this case, we are only using one classifier, so we must ensure that it's enabled and that its confidence threshold is set to 0% so that all results from **Keyword Based Classifier** are returned.

7. Click **Configure Classifiers**. Set the minimum confidence to 0%, and make sure that the checkbox under **Keyboard Based Classifier** is filled:

Figure 6.28 – Setting the Keyword Based Classifier to 0% confidence

With that done, our Classifiers Scope is complete. The next step is checking the confidence of the classifier. In this case, if the confidence returned by our classifier is below a certain percentage, let's have the automation launch the Classification Station to get a more accurate classification from the user.

8. Drag an **If** activity after the Classification Scope, inserting the following as the condition:

```
classification_results(0).Confidence < .5
```

9. Inside of the then block, drag a **Classification Station** activity (*Figure 6.29*). Insert our file path, document text, DOM, taxonomy, and classification results as input. Insert our classification results again as output:

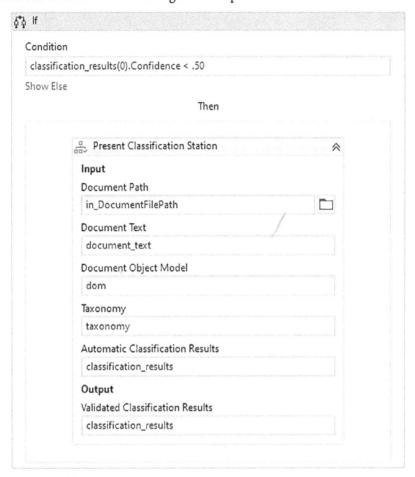

Figure 6.29 – Setting up the Classification Station

In conjunction with the Classification Station, we need a trainer to train the Keyword Based Classifier with our human-validated data.

10. After **Present Classification Station**, insert **Train Classifiers Scope**. Insert our file path, document text, DOM, taxonomy, and classification results (under **Human validated classification**) as input.

11. Drag **Keyword Based Classifier Trainer** into **Train Classifiers Scope**.

12. Click **Manage Learning** and point to our `classification.json` file.

13. Lastly, click on **Configure Classifiers**, selecting **Keyword Based Classifier Trainer** for the **Sliced Invoices** document type (*Figure 6.30*):

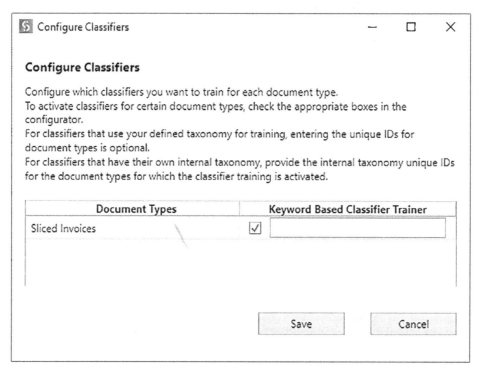

Figure 6.30 – Saving the Keyword Based Classifier Trainer

And that's it for classifying! Try running the automation. If the Classification Station does not appear, then try increasing the confidence threshold in the **If** statement. Try also inserting a *log message* or a *message box* activity to print the confidence, and note how the confidence increases after each validated run.

Now, we have loaded the taxonomy, digitized text, and classified our invoice document. In the next section, we'll learn about extracting data from digitized text.

Introducing extraction

Once the classification of our document types is complete, automation is now ready to extract relevant information from each document. With extraction, automation will extract relevant data fields based on the taxonomy created at the beginning of the Document Understanding framework.

Introducing the Data Extraction Scope

Extraction is performed through the **Data Extraction Scope** activity:

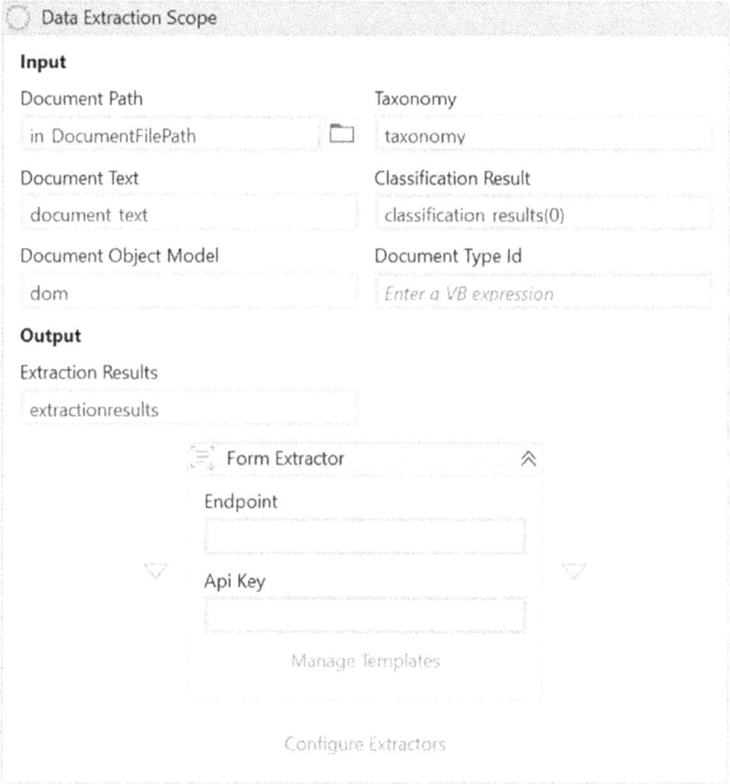

Figure 6.31 – Data Extraction Scope

Within the **Data Extraction Scope** activity, let's cover the **Input** and **Output** parameters present:

- **Input:**

  - **Document Path**: The full path to our target document

  - **Taxonomy**: The taxonomy returned from the Taxonomy Manager

  - **Classification Result**: The results from classification

  - **Document Text**: The document text returned from digitization

  - **Document Type Id**: The document type ID found in Taxonomy Manager – this field is optional if the classification result is used

  - **Document Object Model**: The DOM returned from digitization

- **Output**:

  - **Extraction Results**: The results stored in an `ExtractionResult` object

With the Data Extraction Scope activity introduced, let's continue with the Extraction Wizard within the Data Extraction Scope activity.

Introducing the Extraction Wizard

Like the Classifiers Scope, there is an area to insert different extractors; this allows us to leverage multiple extractors (*Figure 6.32*):

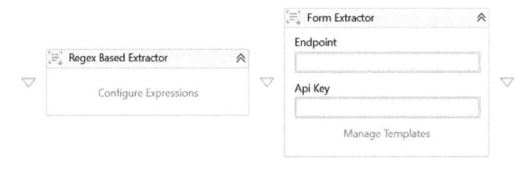

Figure 6.32 – Leveraging multiple extractors at once

However, with the Extraction Wizard, you can configure the use of multiple extractors at the **data field** level. This allows for a hybrid approach (*Figure 6.33*), where an extractor and a minimum threshold can be set at the data field level.

To open the Extraction Wizard, click the **Configure Extractors** button of the Data Extraction Scope. Like the Classifiers Scope, extractors will be executed in the order of priority (left to right). Each extractor will also have a minimum confidence percentage to act as a confidence threshold. One difference from the Classifiers Wizard is **Framework Alias**: this field is used to map an extractor to a trainer. Each extractor has a unique alias, while trainers can share the same alias:

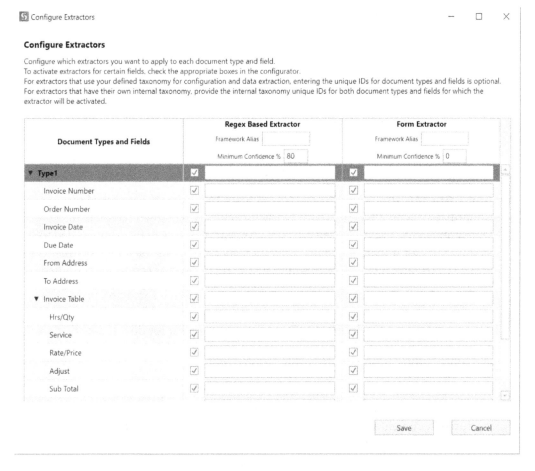

Figure 6.33 – Leveraging multiple classifiers at once

Note within each extractor (**Regex Based Extractor** and **Form Extractor** from *Figure 6.33*) the ability to toggle a checkbox to the right of each field within a document type. This enables a hybrid approach of leveraging multiple classifiers per data field, allowing us to set a specific extractor per data field, or allowing us to set multiple extractors as "fallback" extractors if the highest priority extractor returns low confidence.

Reviewing the types of extractors

Based on the requirements of your use case, you may be drawn to use one or more of the following extractors available in Document Understanding:

- **Regex Based Extractor**: This is a simple extractor that extracts data based on regular expression matches. This extractor is best for simpler use cases, where the targeted data field is in a standard, predictable format. While this extractor can support both text fields and table fields, it cannot be retrained. If there is variability in your target data fields (in format and context), then the Form or Machine Learning extractors may be better choices.

- **Form Extractor**: This is another simple extractor that extracts data based on standardized documents. The Form Extractor relies on a defined template to identify and extract values based on page-level anchors. It is not recommended to use this extractor if there are many different layouts to handle or if the documents are skewed (rotated, curved, or multiple sizes). This extractor cannot be retrained.

- **Intelligent Form Extractor**: This is built on top of the Form Extractor, providing extra features such as handwriting recognition, handwritten data extraction, and signature detection. The Intelligent Form Extractor is great for documents that are printed or handwritten, or for use cases that require validation of a signature. This extractor cannot be retrained.

- **Machine Learning Classifier**: This leverages an ML skill deployed to UiPath AI Center to extract documents. This extractor is great for structured or semi-structured documents of varying layouts. It can be retrained with the *Machine Learning Extractor Trainer*.

- **FlexiCapture Extractor**: This is built by ABBYY to extract documents based on ABBYY FlexiCapture definitions. This extractor can be found within the `UiPath.Abbyy.Activities` package.

We'll discuss further how to use some of these extractors in later sections and chapters.

Validating extractors

Like classification, we may require human validation of extractor output with the use of the **Validation Station**. The use of the Validation Station is generally recommended to enforce 100% accuracy – it is always best to have a human perform a quick check before returning to automation.

The Validation Station is available for both attended and unattended automation through the use of the **Present Validation Station** activity for attended, and the **Create Document Validation Action** and **Wait for Document Validation Action and Resume** activities of UiPath Action Center.

When presented with the Validation Station, the user will have the ability to review and correct extractions performed by automation. In the Validation Station, the user can rotate pages, validate document types, and validate extractions performed by an extractor (*Figure 6.34*):

Figure 6.34 – The Validation Station

With our extraction validation completed with the Validation Station, we can leverage our human validation to retrain and strengthen the classifiers used. In the next section, we'll discuss how to train extractors.

Training extractors

Similar to classification, we can close the feedback loop and retrain our extractors after validation. This retraining model helps improve the performance of our extractors for future executions. Training of classifiers is performed through the **Train Extractors Scope** activity. Within the **Train Extractors Scope** activity, let's cover the **Input** parameters present:

- **Document Path**: The full path to our target document
- **Document Text**: The document text returned from digitization
- **Document Object Model**: The DOM returned from digitization
- **Human Validated Data**: The Validation Station results

The following extractors are retrainable:

- The Machine Learning Classifier

Extracting data from the document

Let's continue our example with extraction.

We'll follow a similar framework as classification for extraction. Let's extract the data and validate whether our confidence is below a threshold:

1. Drag a **Data Extraction Scope** into the workflow.
2. Insert our document file path, taxonomy, document text, and DOM as input. In the **Classification Result** input, insert the 0th classification result, since **Classification Result** is a collection of classification results.

    ```
    classification_results(0)
    ```

3. Create a variable called `extraction_results` (of `ExtractionResult` type) and place it as an output in the Data Extraction Scope.

    For this example, we're going to use the Form Extractor, since we only have one layout with our invoice and its format is not skewed, rotated, or zoomed in any way.

4. Drag **Form Extractor** into **Data Extraction Scope** and insert your API key into the activity.

> **Important Note**
> You can get your API key by navigating to the **Admin > Licenses** page of your **Automation Cloud** account. Click the refresh button next to **Document Understanding** to generate an API key.

5. Click **Manage Templates** within **Form Extractor**, and click **Create Template** to start a new template.

6. When prompted, configure the **Create a new template** form (*Figure 6.35*), using the PDF invoice as a template:

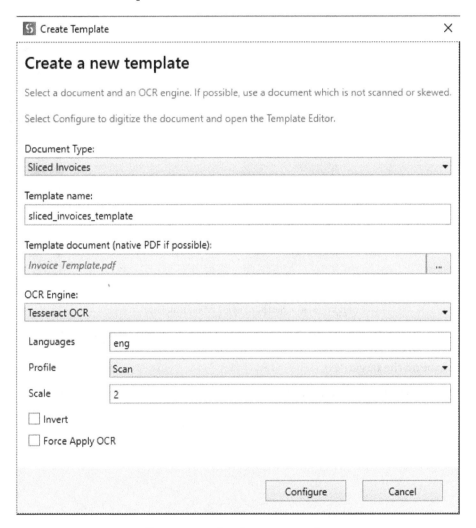

Figure 6.35 – Creating a new template

7.   Once open, you will be prompted to choose **Page 1 Matching Info**. Control + Click **Invoice** at the top-right corner, the **Thanks for choosing…** sentence at the bottom of the page, and **Page 1 of 1** at the bottom of the page:

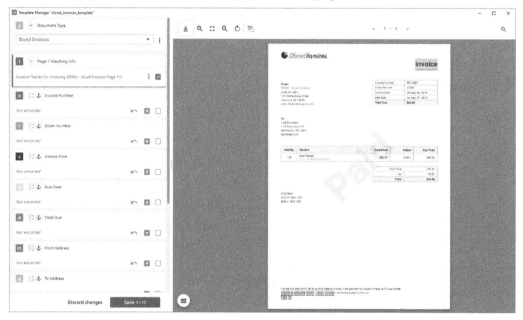

Figure 6.36 – Setting Page 1 Matching Info

8.   For the data fields defined by our taxonomy, choose anchor selection mode (keys *D* + *S*), and then left-click and highlight the relevant text on the form, drawing a blue box around the target text.

9.   Next, choose your anchor(s) by pressing *Ctrl* + clicking on the relevant ones:

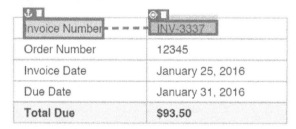

Figure 6.37 – Choosing data fields for Invoice Number

10. Once your text is highlighted, to confirm, click the **Extract Value** plus sign (keys *F + C*) under your data field:

Figure 6.38 – Confirming our selection

11. Perform the same operation for each data field and each data row within the invoice table. Click **Save** to save changes and **Close** to close the Template Manager.

12. Click **Configure Extractors** within the Data Extraction Scope and choose **Form Extractor** for the **Sliced Invoices** document type – we'll be using the Form Extractor for all data fields of the **Sliced Invoices** document type:

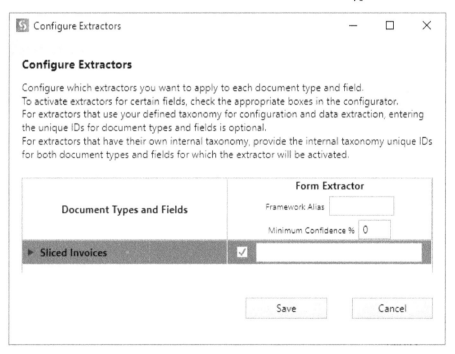

Figure 6.39 – Configure Extractors

Lastly, let's present the Validation Station. The Form Extractor is not retrainable, but we can present the Validation Station to the human user to double-check whether our extraction is correct.

13. After our Data Extraction Scope, drag a **Present Validation Station** activity (*Figure 6.40*). Insert our file path, document text, DOM, taxonomy, and extraction results as input. Insert our extraction results again as output:

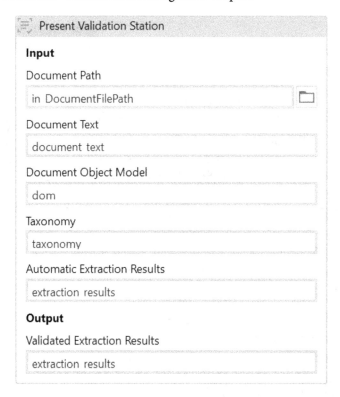

Figure 6.40 – Present Validation Station

And there you have it! We have extracted the relevant data fields from our invoice and presented the user with the Validation Station to double-check our extraction. One last step (exporting) and we'll be complete with this example!

Introducing exportation

The last step of the Document Understanding framework is exporting our extracted and validated data into a data structure that automation can use. With the **Export Extraction Results** activity, we can extract our document data into a `Dataset` variable – a collection of data tables, including a column for each data field defined in the taxonomy, and additional columns for the confidence of each extraction.

With this extracted data, we can easily continue the automation process – whether processing the data directly or inserting it into Excel.

Exporting data from the document

Let's wrap up our example with an export.

Exporting our validated data is very quick; we just need to use the **Export Extraction Results** activity to export our data into a data table, and then we can use that data later in the automation (such as adding to Excel, emailing out, or adding more logic).

1. Drag an **Export Extraction Results** activity to the end of the workflow.

2. Insert our extraction results variable as input.

3. Create a variable called `dataSet` (of the `DataSet` type) and insert it as output:

Figure 6.41 – Export Extraction Results activity

And that's it! Try writing the dataset into Excel or outputting the dataset as a string, and then print it out to see your result!

The Document Understanding framework supplies us with an easy-to-follow list of components for us to build a Document Understanding use case. We learned about the various components of the framework and how to use each. Let's now continue with UiPath's AI Center.

Getting started with UiPath AI Center

UiPath AI Center unleashes the capability to leverage ML models with RPA. With AI Center, we're able to deploy, manage, and retrain out-of-the-box or custom ML models. In this section, we'll navigate around the core concepts of AI Center and dive into the activities of the AI Center activity package.

Using AI Center

In this section, we will navigate around the key concepts of AI Center. By getting familiar with these concepts, we will build experience with leveraging AI Center, key for building automation with ML capabilities.

When creating an automation project, we generally want to follow two paths into production:

- **Use pretrained models**: The developer should follow the following flow:

    I.   Creating a project

    II.   Leveraging an ML package

    III.   Deploying it as an ML skill

    IV.   Dragging and dropping the ML model into RPA

- **Use retrainable models**: The developer should follow the following flow:

    I.   Creating a project

    II.   Uploading a dataset

    III.   Leveraging an ML package

    IV.   Deploying a training pipeline to train the model

    V.   Deploying it as an ML skill

    VI.   Dragging and dropping the ML model into RPA

    VII. Connecting RPA back with the ML model, providing continuous feedback

In this section, we'll investigate the key concepts of AI Center to follow these two paths into production.

Introducing projects

Every AI Center use case starts by creating a project in AI Center. A project is a collection of resources (ML packages, datasets, and pipelines) that you will leverage as you build out your cognitive automation use case.

Creating a project

Let's work together to build a simple text classification project with AI Center, starting off our exercise by creating a project in AI Center:

1. Navigate to your UiPath Cloud, and click on **AI Center** to get started:

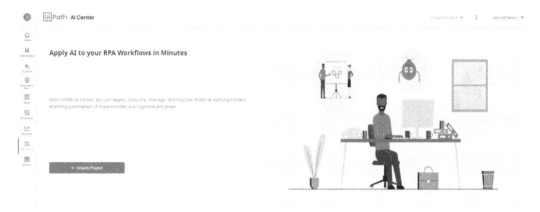

Figure 6.42 – The AI Center home page in the UiPath Cloud

2. Once in AI Center, click **Create Project** and provide a descriptive project name and description.

After providing AI Center with a name and a description, you will be navigated to the project dashboard. Within this project dashboard, you will notice tabs for our datasets, packages, pipelines, and skills, which we'll touch upon in later sections. Let's continue to **Datasets** within AI Center.

Introducing datasets

Now that we have a project created, let's look at the first tab – **Datasets**. By definition, a dataset is merely a set of data. This is no different with AI Center, where a dataset is a collection of files and folders leveraged by an AI Center project. Within **Datasets**, we can upload any set of files and folders necessary to allow our ML models to access data.

Creating a dataset

Let's add our training and testing datasets to AI Center:

1. Within the project dashboard, click **Datasets** to navigate to the **Datasets** page.

2. Click **Create New** to create a new dataset, providing a descriptive name for our dataset.

3. After the dataset is created, click **Upload** to upload the training and testing folders from the GitHub link to AI Center: `https://github.com/ PacktPublishing/Democratizing-Artificial-Intelligence- with-UiPath/tree/main/Chapter06/AI_Center_Test/Amazon_ Reviews_Dataset`.

With our dataset loaded, let's take a look at the ML packages that will leverage our uploaded dataset.

Introducing packages

Similar to how a package in Orchestrator can have various workflow, text, and Excel files, an ML package in AI Center is a folder containing all the code and associated files necessary to serve an ML model:

Figure 6.43 – AI Center ML packages

Within the **ML Packages** option of the project dashboard, you will notice two options (*Figure 6.43*):

-   **Upload zip file**: The ability to upload custom ML packages built in Python

-   **Out of the box Packages**: The ability to use UiPath-developed or open source ML packages

Reviewing out-of-the-box packages

As described in earlier sections, AI Center provides you with many UiPath and open source ML packages, allowing you to quickly get up and running with cognitive automation. The following out-of-the-box packages are available:

- UiPath Document Understanding
- UiPath Image Analysis
- UiPath Language Analysis
- UiPath Task Mining
- Open source Image Analysis
- Open source Language Analysis
- Open source Language Comprehension
- Open source Language Translation
- Open source Tabular Data
- Open source Time Series

With the out-of-the-box packages reviewed, let's look into creating a package next.

Creating a package

Because we're performing a text classification exercise, let's choose UiPath's out-of-the-box ML package:

1. Navigate to ML Packages by clicking the **ML Packages** tab in the project dashboard.
2. When presented with **Create a new package**, click **Out of the box Packages**.
3. Underneath the out-of-the-box packages, choose **Language Analysis** > **EnglishTextClassification**, choosing **Package Version 6.0**.

After providing the ML package with a name, you should see your new ML package in your project dashboard, as shown in *Figure 6.44*:

Figure 6.44 – A newly created TextClassification ML package

With our package created, let's investigate creating a pipeline that can train our deployed ML package.

Introducing pipelines

With our ML package selected, we can now venture into ML pipelines. An ML pipeline is a series of steps taken to produce an ML model. During the pipeline process, we will provide input, and once completed, the pipeline will produce a set of useable output and logs.

In AI Center, there are three types of pipelines:

- **A training pipeline**: Produces a newly trained ML package version from an input ML package and dataset

- **An evaluation pipeline**: Produces a set of metrics and logs from a newly trained ML package version and dataset

- **A full pipeline**: A full data process pipeline, performing both training and evaluation pipelines

Creating a pipeline

For this exercise, we will use our training and testing datasets to run a training and evaluation pipeline:

1. Navigate to **Pipelines** by clicking the **Pipelines** tab in the project dashboard.

2. Click **Create New** to create a new pipeline, selecting the following:

   - **Pipeline type**: **Train run**

   - **Choose package**: **TextClassification**

- **Choose package major version: 6**
- **Choose package minor version: 0**
- **Choose input dataset**: `Amazon Reviews/train/`
- **Enable GPU**: No:

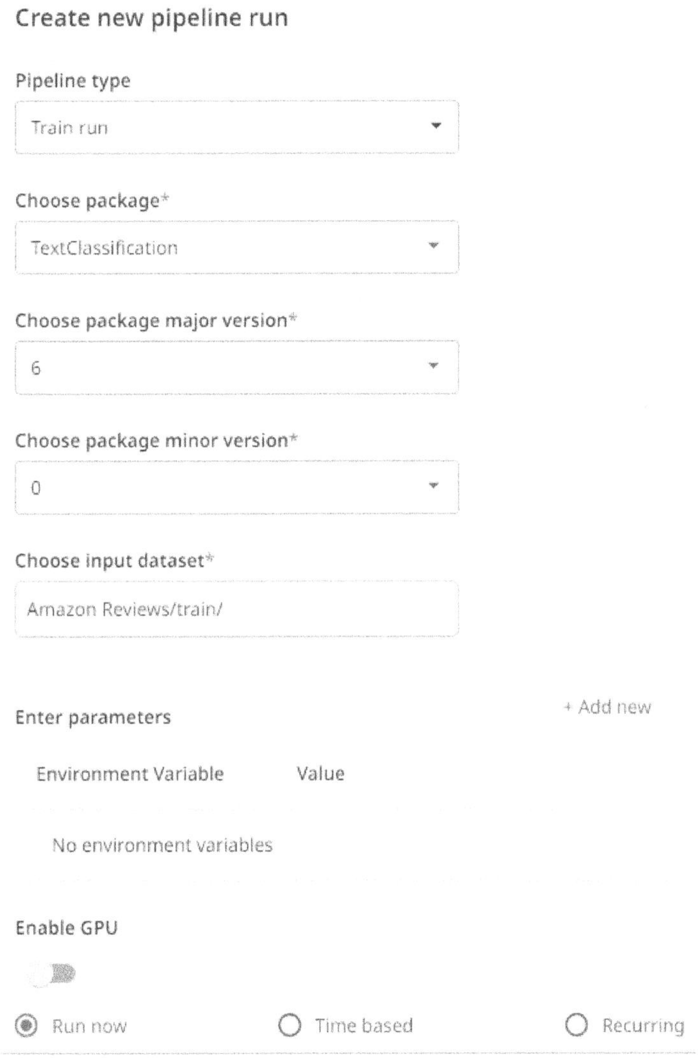

Create new pipeline run

Pipeline type

Train run

Choose package*

TextClassification

Choose package major version*

6

Choose package minor version*

0

Choose input dataset*

Amazon Reviews/train/

Enter parameters   + Add new

Environment Variable   Value

No environment variables

Enable GPU

◉ Run now   ○ Time based   ○ Recurring

Figure 6.45 – Creating the training pipeline

3. Click **Create** to start the training pipeline.

> **Important Note**
> The size of a dataset will affect pipeline duration time. Choosing **Enable GPU** can lead to quicker output, but GPUs are not available with UiPath Community edition.

4. Perform steps *2* and *3* again, choosing **Evaluation run** as the pipeline type, version 6.1 of the Test Classification package, and the `test` dataset folder as the input dataset:

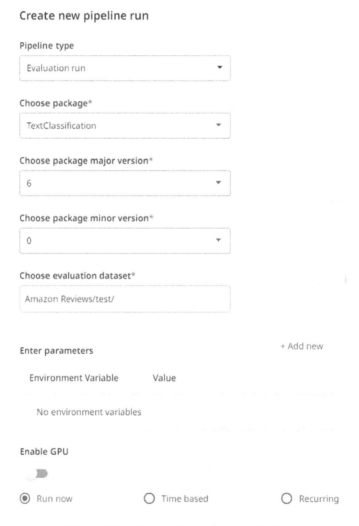

Figure 6.46 – Creating the evaluation pipeline

Once completed, the evaluation pipeline will return a few artifacts as output:

- `Evaluation.csv`: Data used to evaluate the model
- `Evaluation-report.pdf`: An output report of the pipeline run

Clicking on the evaluation pipeline on the **Pipelines** page will showcase the pipeline's output and logs:

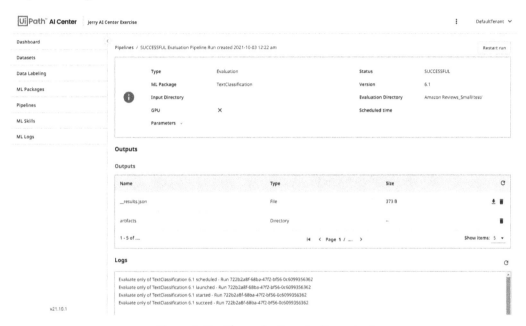

Figure 6.47 – The evaluation pipeline output

And that's it! In the next section, we'll deploy an ML skill and attach it to an RPA workflow.

Introducing ML skills

Similar to how a package is a deployed RPA project from UiPath Studio, an ML skill is a deployed ML package from AI Center.

Once deployed into AI Center, an ML skill can be consumed by automation. Choosing the ML skill activity in UiPath Studio, you can select from a list of ML skills deployed on AI Fabric, passing input data and leveraging output data as needed:

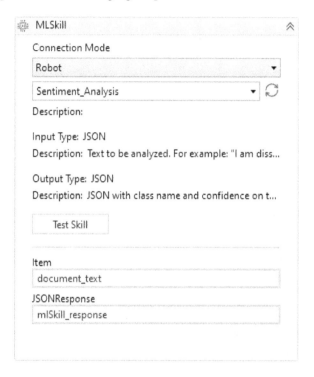

Figure 6.48 – The MLSkill activity in UIPath Studio connected to an ML skill from AI Center

> **Important Note**
>
> Consuming an ML skill in UiPath Studio requires UiPath Studio v2019.10+ with UiPath Robot v2019.10+ and the `UiPath.MLServices.Activities` package installed.

Creating an ML skill

With our model trained and evaluated, we can now deploy it into a skill usable with RPA:

1. Navigate to **ML Skills** by clicking the **ML Skills** tab in the project dashboard.
2. Click **Create New** to create a new ML skill, selecting the following:

    - **Name**: `Amazon_Reviews`
    - **Choose package**: **TextClassification**
    - **Choose package major version**: **6**
    - **Choose package minor version**: **1**:

Figure 6.49 – Creating an ML skill

3. Click **Create** to create an ML skill. The skill will take some time to package before becoming available for use.

4. Start a new project in UiPath Studio and open `Main.xaml`.

> **Important Note**
> Before starting with this exercise, make sure you have the following installed: `UiPath.MLServices.Activites` v1.1.8 or higher.

5. Inside `Main.xaml`, create a string variable, `amazon_review`.

6. Drag an Assign activity, setting the **To** field as `amazon_review` and the **Value** field as the following:

```
"I love this product!"
```

7. Drag an ML Skill activity, choosing our `Amazon_Reviews` skill, and assign the **Item** input as our `amazon_review` string and create a `json_response` variable as the **JSONResponse** output:

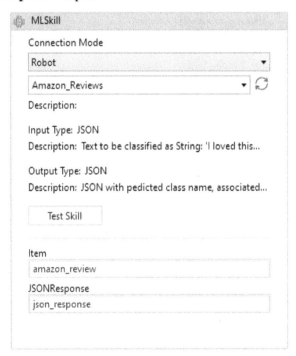

Figure 6.50 – The Amazon_Reviews ML skill

8. After the ML skill activity, drag a Message Box activity, placing the `json_response` variable as input.

Try running it! Note the JSON response in the message box. We can replicate this type of testing by clicking **Test Skill** within the ML Skill activity. In later exercises, we will parse this JSON as part of a further solution, but for now, we have a working ML skill trained and connected to RPA! In the next section, we will investigate AI Center logs.

Introducing logs

The logs page of AI Center is a consolidated view of all events occurring in AI Center, such as the logs page of Orchestrator.

The following are events logged within AI Center:

- **ML package validation events**: When a model is updated
- **Pipeline events**: When a pipeline starts or fails
- **ML skill deployment events**: When a skill is created
- **ML skill prediction events**: When a skill is working, throwing a failure:

Figure 6.51 – ML Logs examples

We learned about the various components of AI Center, and how to use each. In subsequent chapters, we'll use AI Center as we build use case examples, but for now, let's continue with UiPath's Computer Vision.

Getting started with UiPath Computer Vision

UiPath Computer Vision gives automation an additional capability to see a computer screen and identify elements on the page versus the traditional spying of elements through sets of properties and attributes. This unleashes the capability to intelligently interact with more complex use cases that include VDIs, unsupported applications, and even operating systems.

To interact with elements (such as text fields and buttons) on a page, traditional RPA automation leverages properties or attributes of these elements, called **selectors**. UiPath automation is typically compatible with most applications and able to find selectors for automation to interact with; however, there exist some technologies for which finding selectors is unfeasible, and alternate solutions are necessary.

In the past, most use cases that contained applications without findable selectors were either descoped or solved with brittle solutions such as image recognition. In order to help bridge the gap and provide a solution to these types of use cases, UiPath developed Computer Vision, an ML-based method used to locate elements on a screen. This has opened up a new spectrum of automation opportunities that can be developed, as automation can now interact on user interfaces that were not automatable with extensions such as the UiPath Citrix or VMware extensions, or interfaces that were generally just too complex to interact with.

In this section, we'll learn about UiPath's Computer Vision, how to use it, and its available activities.

Using Computer Vision

Before using the Computer Vision activities, ensure you have the UI Automation package 19.10 or greater installed. An API key from Automation Cloud must also be generated:

1. Navigate to the **Admin** > **Licenses** page of **Automation Cloud**.

2. Click the refresh button next to **Computer Vision** to generate an API Key:

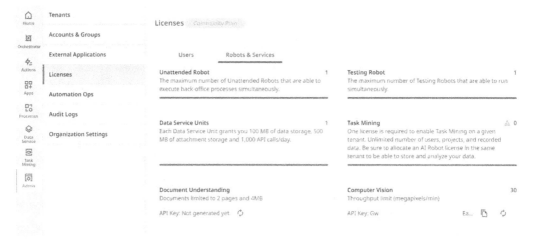

Figure 6.52 – The Computer Vision API key through Automation Cloud

Once an API key is generated, it must be added to the project settings of your automation project:

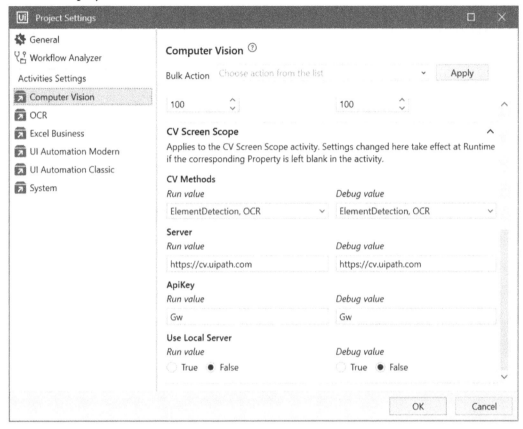

Figure 6.53 – Adding your API key to Project Settings

The **Computer Vision (CV)** package contains similar UI Automation activities used for traditional interface automation, such as **Click, Type Into, or Get Text**. The main difference between the CV activities is the use of the CV Screen Scope. Once a CV Screen Scope is placed into the workflow and a screen is targeted, you can use all the CV activities to build your automation, such as the following:

- CV Click
- CV Element Exists
- CV Get Text

Figure 6.54 – CV Get Text within a CV Screen Scope

- CV Highlight
- CV Hover
- CV Type Into

UiPath CV gives us the additional capability to see a computer screen and identify elements on a page when the traditional spying of elements doesn't provide consistent results.

In this section, we learned about UiPath CV, what it is, and how to use it.

Summary

In this chapter, we covered the development milestone of the automation development life cycle by diving into further understanding the cognitive tools provided by UiPath. We learned about the basics of Document Understanding, AI Center, and Computer Vision to understand how we can tie each tool into an automation workflow.

The tools we reviewed in this chapter are the building blocks for cognitive automation. By learning about the basics of each of these tools, we now have added tools to our automation toolkit that we can use in future use cases. This introductory understanding of these tools will help us in subsequent sections where we leverage them again to build actual use cases.

In *Chapter 7, Testing and Refining Development Efforts*, we will take the next step with cognitive automation opportunities by learning about how to prepare a cognitive automation opportunity for testing and UAT. In the next chapter, we will learn how to gather test data and create test cases for UAT, in addition to closing the feedback loop with AI Center.

QnA

1. What are the characteristics of the Document Understanding framework?

   - Taxonomy, digitization, classification, extraction, and export.

2. Can you digitize a document without an OCR engine?

   - Yes, you can; however, the digitization activity of the Document Understanding framework will require an OCR engine, even if OCR is unneeded.

3. When should the Classification Station or Validation Station be used?

    - They should be used when automation confidence is not high and human validation is needed to augment automation. They should also be used when closing the feedback loop and retraining via human validation.

4. Is AI Center limited to out-of-the-box ML packages?

    - No. AI Center can handle custom ML models made with Python in addition to the out-of-the-box packages it offers.

5. When is CV best used?

    - CV is best for applications that contain unreliable selectors or VDI/VM environments where selectors cannot be captured.

7

Testing and Refining Development Efforts

Accurately and comprehensively testing automation prior to deployment is paramount in ensuring the longevity of the automation. By creating a proper testing approach to gather test data, create test cases, and execute tests, we can ensure full coverage when assessing automation development.

In this chapter, we will review the approach to testing cognitive automation, how to prepare test data and test cases for **user acceptance testing** (**UAT**), and how to train and increase model accuracy by closing the feedback loop with human validation and UiPath's built-in validation features.

In this chapter, we're going to cover the following main topics:

- Approaching cognitive automation testing
- Executing cognitive automation testing
- Closing the feedback loop

Approaching cognitive automation testing

As we continue through the automation life cycle once we have developed our automation, we can continue to test what was developed. While adding a cognitive component (Document Understanding, **machine learning (ML)** skills, and so on) brings additional complexities to automation from traditional rules-based **robotic process automation (RPA)**, testing cognitive automation isn't too different from the traditional approach to testing automation.

In *Chapter 6*, *Understanding Your Tools*, we saw that when developing cognitive automation, we were essentially developing two distinct components: **RPA** and **cognitive automation**—we would build the ML skill first, then build RPA to leverage that ML skill. With this mindset, we can think of RPA and cognitive automation as modular components of the broader automation project, mimicking the design principle called **separation of concerns (SoC)**.

> SoC
>
> This is a software design pattern where we separate a computer program into distinct sections.

In the case of **cognitive automation**, the RPA component only needs to be prepared to handle the outputs returned from the cognitive component, and our cognitive component only needs to be prepared to handle inputs it receives from the RPA component. With that in mind, we can test these expectations while also splitting the testing approach because these components (or sections) are separate, as illustrated in the following diagram:

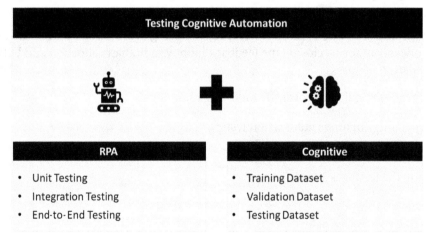

Figure 7.1 – Testing cognitive automation

As shown in *Figure 7.1*, we can approach testing cognitive automation in two parts, leveraging traditional RPA testing approaches and ML test approaches to each component. We'll proceed as follows:

- **Testing RPA**: We will approach testing RPA in the same manner that we test any RPA project, through a mixture of unit testing, integration testing, **end-to-end (E2E)** testing, and so on.

- **Testing cognitive automation**: We will approach testing cognitive automation in a similar manner to how we test an ML model, through data-driven training, evaluating, and testing.

Once we successfully test both RPA and cognitive automation in their own environments, we can bring both components together for UAT prior to deploying into production.

Let's take a look at how we can test RPA and cognitive development.

How to test RPA development

Testing RPA development in a comprehensive manner is pivotal to gaining the confidence of the business and building a successful automation program. However, there are many components of RPA that could lead to exceptions, such as the following:

- **Application issues**: Technical changes; customizations

- **Automation issues**: Business logic changes

- **Environment issues**: System updates; desktop changes

Thus, when testing RPA, we need to take a prescriptive approach to ensure reliability.

Because RPA doesn't deviate much from traditional programming, we can leverage traditional software development testing practices when testing automation. In software testing, there are many different types of testing leveraged to ensure code is performing as desired, as illustrated in the following diagram:

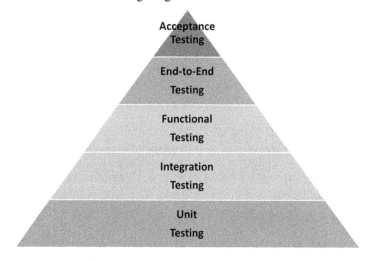

Figure 7.2 – Hierarchy of development testing

The following list covers the hierarchy in brief:

- **Unit Testing**: Lowest level of testing—tests individual code components or functions used by software

- **Integration Testing**: Verifies different modules or components work together

- **Functional Testing**: Verifies business requirements, such as verifying the output of an action or logic is as expected

- **End-to-End Testing**: Replicates user behavior and verifies theE2E flow is running as expected

- **Acceptance Testing**: Series of formal tests to ensure development matches business requirements

For more information on the types of software testing, you can refer to the following link:

```
https://www.atlassian.com/continuous-delivery/software-
testing/types-of-software-testing
```

While automation programs may have different levels of formal testing required during their automation life cycles, all development must undergo some levels of formal acceptance testing. This leaves the lower levels of testing (unit to E2E) to the development team to perform prior to UAT.

How to test cognitive components

When testing a cognitive component, all testing starts with our data. In *Chapter 6, Understanding Your Tools*, we completed an AI Center example with a training and evaluation dataset and pipeline. However, as we approach deployment of an ML skill, we need to add another dataset—the training dataset, leaving us with the following datasets necessary when building cognitive automation:

Figure 7.3 – Testing an ML skill

These datasets are outlined in more detail here:

- **Training dataset**: Initial dataset to train the ML model. This training set should simulate our ideal expected result—or *ground truth*, in other words.

- **Evaluation dataset**: A smaller set of data used to evaluate the training of the ML model.

- **Testing dataset**: An unbiased dataset used to expose the trained ML model to real-life data.

> **Important Note**
> For more information about **ground truth**, check out the following link:
> ```
> https://towardsdatascience.com/in-ai-the-
> objective-is-subjective-4614795d179b
> ```

In summary, we can see that RPA and cognitive components require different types of testing. RPA requires more of a traditional **software development life cycle (SDLC)** testing approach, while cognitive automation requires an iterative, data-driven approach. While respecting the mindset of **SoC**, we can effectively silo development and testing of both RPA and cognitive automation to ensure both components work correctly before bridging them together for UAT. In the next section, we will venture into executing cognitive automation testing.

Executing cognitive automation testing

For successful testing and deployment of cognitive automation, we should always prepare an approach to testing. In this section, let's review details on gathering test data, executing RPA and cognitive tests, and tying it all together with executing UAT tests.

Gathering test data

ML depends highly on data. Having the right types and the right amount of data is crucial in having a successful ML model deployed; therefore, data preparation is such an important part of the ML process.

When gathering test data, many organizations ask how much data is necessary to get started. Unfortunately, there isn't an explicit answer to this question, as there are many variables that can affect how much data is necessary, such as the following:

- The complexity of the business problem ML must solve
- The number of classifications (if necessary)
- The complexity of the algorithm used

If necessary, you can try to target the following estimations when gathering data. Of course, every problem may be slightly different, thus more data may be required to output satisfactory results:

- **Text classification**: Try to have a couple of hundred examples of *EACH* classification.
- **Tree-based Pipeline Optimization Tool (TPOT) tabular classification**: Again, a few hundred examples of each classification.

In general, when trying to gather data from a business team, you want to strive for as many data points as possible. It's always easier to have too much data and to cut down, versus going back to the business and requesting more data.

After you gather the data necessary for the project, it is important to have enough data to train, validate, and test the ML model, as shown in the following screenshot:

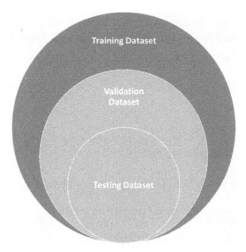

Figure 7.4 – Splitting our dataset

A majority of our dataset should be used for training the ML model, while smaller datasets can be used to validate the model and to finally test the model. Here's an estimated rule of thumb:

- **Training Dataset**: 60-65%
- **Validation Dataset**: 20-25%
- **Testing Dataset**: 10-20%

Once our data for executing testing is gathered, we can continue on to the next section of cognitive automation testing: execution.

Executing RPA testing

As we test RPA, we should start with the hierarchy of development testing (*Figure 7.2*) from earlier in the chapter. In order to easily enable this hierarchy, there are a few practices we should follow during development, as outlined here:

- **Small workflows**: Workflows should have a distinct input and output, performing a set action (such as *Log in to X, Navigate to Y, Read Z*).

- **Independent workflows**: A workflow (minus `Main.xaml` or `Process.xaml`) should not call other workflows.

By creating smaller, modular workflows, we can more easily create unit tests to test the functionality of development.

When creating test cases for unit testing, we want to follow a methodology that allows us to prepare our environment to run the workflow, test our workflow, and validate the workflow's output. The methodology is known as the *Given-When-Then* formula of testing and is elaborated on here:

- **Given** some context
- **When** some action is carried out
- **Then** some output should be obtained

For more information on *Given-When-Then*, you can refer to the following link:

`https://www.agilealliance.org/glossary/gwt/`

As test cases are built out in UiPath, separate sequences are used to separate the *Given*, *When* and *Then* portions of the *Given-When-Then* formula, as illustrated in the following diagram:

Figure 7.5 – Creating a test case

In the **Given** section of our formula, we need to prepare any test data and prepare our applications for the workflow we want to test. We then do the following:

- Inside of this sequence, we can invoke the `InitAllSettings.xaml` file of our **robotic enterprise framework** (**REFramework**) to load any assets, but also open an application and log in to prepare for the test.

Next, in the **When** section of our formula, we just need to invoke the modular workflow that we want to test.

Lastly, in the **Then** section of our formula, we check the result of what was performed in the **When** section.

> **Note**
>
> Inside of this sequence, we can check to see if a target application is in the correct window or if an output from the workflow invoked in the **Then** section is correct.

Every modular workflow created should have an associated test case created. By following the *Given-Then-When* formula, we now have test cases that not only test our development but can be used in the future to regression test.

With these test cases, we must test each case, ensuring success before moving development further in the development life cycle. To help us test each test case, we can leverage REFramework's test framework to run each of our test cases and report results, as illustrated in the following screenshot:

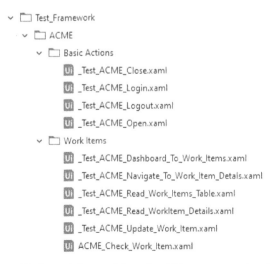

Figure 7.6 – Each modular workflow should have an associated test case

With REFramework's `Test_Framework` folder, we can run each test case created using the `Run_Tests.xaml` file, outputting the results of our test into an Excel and TXT log file, as shown in the following screenshot.

WorkflowFile	Exception	Status	Comments
Framework\InitAllSettings.xaml	Success	PASS	
Framework\InitAllSettings.xaml	Success	PASS	
Test_Framework\ACME\Basic Actions_Test_ACME_Open.xaml	Success	PASS	
Test_Framework\ACME\Basic Actions_Test_ACME_Login.xaml	Success	FAIL	AppEx: Account not found: John_Doe@xyz.com at Source: ACME Login: Send Outlook Mail Message
Test_Framework\ACME\Basic Actions_Test_ACME_Logout.xaml	Success	PASS	
Test_Framework\ACME\Basic Actions_Test_ACME_Close.xaml	Success	PASS	
Test_Framework\SHA1\Main Page_Test_SHA1_Read_Hash.xaml	Success	PASS	
Test_Framework\SHA1\Main Page_Test_SHA1_Enter_Data.xaml	Success	PASS	

Figure 7.7 – Output of REFramework's testing framework

REFramework's testing framework is just one method of testing our development. Other methods include leveraging UiPath's Test Suite to run and report on our test cases. Either way, by following this *Given-Then-When* type framework, we have a method to create test cases for each workflow that we can use for development testing and regression testing post-deployment.

Executing cognitive testing

As we test an ML skill, we should start with the training dataset, building a training pipeline to train the model with an initial set of data. Once trained, the model can be evaluated with the evaluation (or sometimes called validation) dataset. Once our dataset is evaluated, we can review the model's evaluation report to judge the model's performance. This evaluation report is usually a direct output of an evaluation pipeline, as illustrated in the following screenshot:

Evaluation Statistics

accuracy	0.53353
precision	0.52848
recall	0.53353
matthews correlation	0.41685
fscore	0.5275
auroc	0.8411

Figure 7.8 – Example evaluation report

Figure 7.8 is an example of an evaluation of a text classification model. We generally want to aim for a greater-than-70% Fscore. In this case, we're lower than 70%, so we should review our training and evaluation datasets for correctness and investigate editing model parameters or choosing a different model altogether.

Once we choose a model and successfully evaluate it, we can move into testing the model with the testing dataset. This dataset should be completely unbiased and full of new data the model hasn't seen before. The best type of data for testing is young data, preferably coming from our production environment.

Executing UAT

In previous sections, we discussed testing RPA and ML in their own, distinct silos. This works well during the development phases of the automation life cycle, but there should be a convergence of the two technologies before deployment into production. This convergence should occur during the UAT phase of the automation life cycle.

UAT is a type of testing performed to get final approval by the end users or businesses on the performance of the automation. This testing is performed prior to deploying the automation into production, showing the end user how the automation should perform given a set of scenarios. During this testing, we do not focus on specific errors such as selectors but focus on the E2E process flow, ensuring that the automation's output meets the requirements of the end user/business.

The following diagram depicts how UAT fits into the automation life cycle:

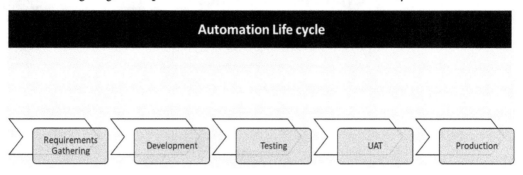

Figure 7.9 – UAT within the automation life cycle

For us to execute UAT, there are a few entry points to consider as criteria prior to final testing. These entry criteria include the following:

- **Development testing completed**: Automation development should be tested (unit, integration, and system tested) and should have run through E2E testing prior to acceptance testing.

- **Peer review completed**: Automation development should be peer-reviewed by another developer for best practice.

- **Environment prepared**: UAT generally occurs in a separate environment from development. This environment can include UAT applications with testing data and a UAT UiRobot.

- **Automation migrated**: Automation should be published into the development orchestrator, committed to source control, and migrated into the UAT orchestrator prior to UAT testing.

- **UAT test plan created**: A test plan with all necessary test scenarios and expected outcomes should be created by the process owner or **subject-matter expert** (**SME**).

- **UAT data gathered**: In conjunction with the test plan, data for UAT should be gathered to test with. Generally, recent production data is best to perform UAT with.

For more information on UAT, you can refer to the following link:

`https://www.guru99.com/user-acceptance-testing.html`

Once all the entry criteria are completed, UAT testing can commence. Executing UAT can take a few forms, depending on whether the business runs the UAT tests or the development team runs the UAT tests for the business to review. An easy way to conduct UAT is to have the SME and development team in a meeting together to run through each test scenario. The SME can supervise the test, while the developer can launch the automation from the UAT orchestrator for each test scenario.

It is important to have exit criteria defined for the completion of UAT. Exit criteria are best defined within a UAT test plan document so that all parties involved know the required criteria for UAT completion. You can see an example of a UAT test plan document in the following screenshot:

#	Scenario	Expected outcome	Case iD	Status
1	* Market Data Rejects > 5% of total integrated * Market_Data_Bypass **is not** true	The robot: * References Market Data portion of output email * Sends Email: - To: team@company.com - Subject: UIP Alert: Market Data Rejects Over 5% - Body: Today's Kyriba Staus Report shows a high amount of Rejections. * END PROCESS	1	APPROVED
2	* Market Data Rejects > 5% of total integrated * Market_Data_Bypass **is** true	The robot: * References Market Data portion of output email * All other functions as expected; formats and distributes email as expected	2	APPROVED
3	* Mark to Market Report **Missing**	The robot: * References Market Data portion of output email * Filters for 'mtm' template * Runs MTM Report using RPA _ FX MTM Generation template * All other functions as expected; formats and distrubutes email as expected	3	APPROVED
4	* Mark to Market Report is 'Warning'	The robot: * References Market Data portion of output email * Filters for 'mtm' template * Sends Email: - To: team@company.com - Subject: UIP Alert Check MTM Status - Body: 'Mark to Market status is: Warning - please investigate' * END PROCESS	4	APPROVED
5	* Mark to Market Report is 'Error'	The robot: * References Market Data portion of output email * Filters for 'mtm' template * Sends Email: - To: team@company.com - Subject: UIP Alert Check MTM Status - Body: 'Mark to Market status is: Error - please investigate' * END PROCESS	5	APPROVED

Figure 7.10 – Example UAT test plan document

Exit criteria can depend on the organization, but generally, UAT exit criteria include the following:

- No defects open
- Successful completion of UAT test scenarios
- Documented sign-off of UAT

To meet these exit criteria, the business analyst, UAT testers, or SME should be documenting the result of each test scenario completed. Depending on the organization, each test scenario can be documented through a screenshot, video recording, or just sign-off on the automation output. Once the exit criteria are met, UAT can be fully completed, and the automation can be deployed into production.

Closing the feedback loop

As cognitive automation is moved from UAT into production, it can face data not seen from the training, evaluation, and testing test sets. We expect the deployed model to be able to handle new data based on the training performed, but there may be times where the model returns an unsatisfactory result or there are opportunities to further improve the model's performance based on the data it encounters.

This is where closing the feedback loop can play a large factor in the performance of an ML model. By closing the feedback loop on a Document Understanding or AI Center ML skill, we can capture unseen data points, using a human to send feedback to the ML skill and continuously train the skill with new data. You can see a representation of closing the feedback loop in the following screenshot:

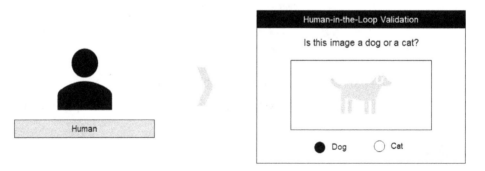

Figure 7.11 – Closing the feedback loop

With UiPath, developers can use a confidence threshold to allow automation to continue with the ML skill's output or prompt users for input by using Attended Automation (Validation Station), or by using Action Center. This feedback loop from human input now allows us to continuously test and refine our model with fresh, relevant data.

In subsequent chapters, we'll dive into how we can close the feedback look with UiPath Validation Station.

Summary

In this chapter, we covered the testing milestone of the automation development life cycle by diving into the approach for testing cognitive automation with UiPath. We learned about the approach to testing cognitive automation, how to prepare for UAT, and how to increase model accuracy with human validation using UiPath's built-in validation features.

In the next chapter, *Receipt Processing with Document Understanding*, we will take a more hands-on approach by working together to build the first of three use cases with UiPath. In the next chapter, we will learn how to use UiPath's Document Understanding to process invoices.

QnA

1. There can be many components of RPA that could lead to exceptions. Which types of issues could arise from automation development?

    -   Application issues, automation issues, environment issues

2. Which types of testing can we perform to test automation?

    -   Unit testing, integration testing, functional testing, E2E testing, acceptance testing

3. What are the three types of datasets necessary for cognitive testing?

    -   Training, evaluation, testing

Section 3:
Building with UiPath Document Understanding, AI Center, and Druid

In this section, we will build practical use cases using UiPath Document Understanding, AI Center, and Druid connections to tie UiPath Automation to cognitive abilities.

This section comprises the following chapters:

8
Use Case 1 – Receipt Processing with Document Understanding

Every organization, irrespective of the industry, has a team that requires receipt processing. Examples of receipt processing can be seen in procurement teams to pay for services delivered, or in HR teams to review expenses submitted by colleagues during business travel. While the use for receipt processing may differ between these teams, one similarity is that processing the receipts is usually a manual and potentially error-prone task to perform. Fortunately, with UiPath's Document Understanding, we can create cognitive automation to apply **Optical Character Recognition (OCR)** to read, classify, and interpret receipts to eliminate manual work.

By having a full understanding of Document Understanding and the *Document Understanding framework*, we can work together to build cognitive automation that can interpret the images of receipts. As we build the use case, we will follow a similar development life cycle to that outlined in the previous chapters. We will start with understanding the current state of our automation opportunity and then design the future state of the process with automation involved. Once the future state is designed, we will build the automation with Document Understanding, test the solution to ensure stability, and finally, deploy the automation into production.

In this chapter, we will cover the following topics as we build automation to process receipts:

- Understanding the current state
- Creating the future state
- Building the solution with the Document Understanding framework
- Testing to ensure stability and improve accuracy
- Deploying with the end user experience in mind

Technical requirements

All code examples for this chapter can be found on GitHub at `https://github.com/PacktPublishing/Democratizing-Artificial-Intelligence-with-UiPath/tree/main/Chapter08`.

Working with UiPath is very easy, and with the Community version, we can get started for free. However, for UiPath AI Center, we will need an enterprise license. You can acquire a 60-day enterprise trial license from `uipath.com`.

With the 60-day enterprise trial, we will have access to the following:

- Five RPA Developer Pro licenses – named user licenses include access to Studio, StudioX, Attended Robot, Apps, Action Center, and Task Capture
- Five unattended robots, five testing robots, and two AI robots
- AI Center, AI Computer Vision, Automation Hub, Data Service, Document Understanding, and Insights

For this chapter, you will require the following:

- UiPath Enterprise Cloud (with AI Center)
- UiPath Studio 2021.4+

- The `UiPath.IntelligentOCR.Activites` package v.4.13.2 or higher
- `UiPath.DocumentUnderstanding.ML.Activites` v1.7.0 or higher

> **Important Note**
>
> Directions on how to install UiPath packages can be found at `https://docs.uipath.com/studio/docs/managing-activities-packages`.

Check out the following video to see the Code In Action at: `https://bit.ly/3J8gX53`

Enabling AI Center in the UiPath Enterprise trial

Additional technical requirements include leveraging AI Center. UiPath Document Understanding comes out of the box with the UiPath Community and Enterprise versions; however, with UiPath AI Center, we need to enable the service within the Enterprise trial. You can enable the service by following these steps:

1. Navigate to **Automation Cloud** at `https://cloud.uipath.com`:

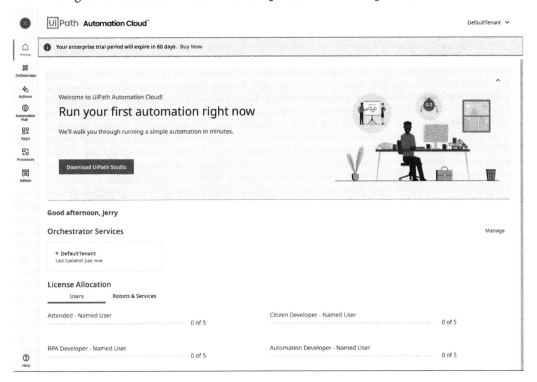

Figure 8.1 – The UiPath Automation Cloud home page

2.   Navigate to **Admin**:

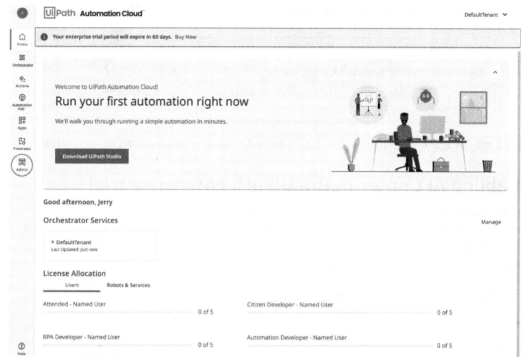

Figure 8.2 – Click on Admin

3.   In **DefaultTenant**, click on the three dots, and then click on **Tenant Settings**:

Figure 8.3 – Click on Tenant Settings

4. Choose **AI Center** under **Provision Services**, and click on **Save**:

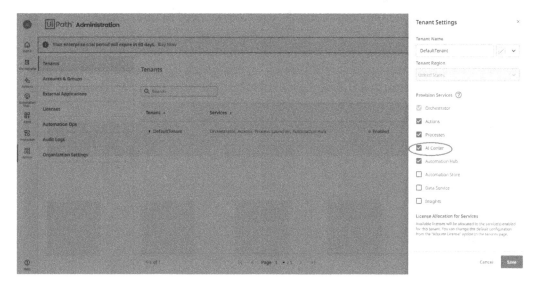

Figure 8.4 – Check AI Center and then click Save

Once AI Center is enabled, we have completed the technical requirements for this chapter. In the next section, we will start gathering context about the use case by reviewing the current state.

Understanding the current state

Before we jump into designing or building the use case, we need to understand the current state of the process. Every day, the **Time and Expense (T&E)** team of Company ABC needs to manually review receipts submitted from the business travels of their colleagues. As the T&E team reviews these receipts, they check them against their expense management tool to ensure that the employee submitted the expense report correctly. The following are the rules set by the T&E team (*Figure 8.5*):

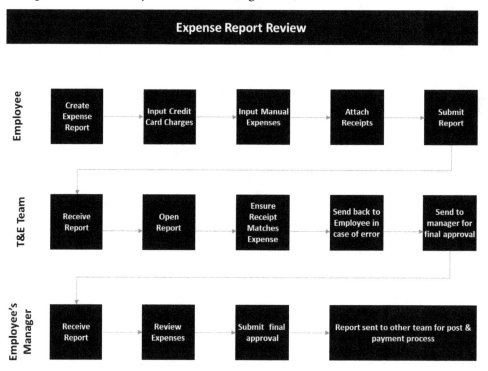

Figure 8.5 – The high-level process flow

Reviewing the high-level process flow, the T&E team automatically receives expense reports once they are submitted by the employee. It is then up to the T&E analyst to open the report and check each expense and receipt for completeness and accuracy. If the report is accurate, the analyst can forward the report to the employee's manager for final approval.

Company ABC has a lot of employees that need to regularly travel, and the company has recently been hiring more employees, so the number of expense reports and receipts that the T&E team has to review is ever increasing. The T&E team has a turnaround time of 2 days, meaning they can review and approve an expense report 2 days after it has been submitted by an employee.

Looking at the current state, there appears to be plenty of opportunity for automation – reverting to the characteristics of an automation opportunity from *Chapter 4*, *Identifying Cognitive Opportunities*, we can see that this automation opportunity has the following characteristics:

- **Manual interactions with applications**: T&E analysts have to manually interact with the expense management tool.

- **Repetitive in nature**: T&E analysts have to perform the same series of steps for each expense and receipt.

- **Is time-consuming**: Reviewing individual receipts can be time-consuming for an analyst to perform.

- **Is rule-based and structured**: The rules surrounding expense reports are mostly structured.

The only area of concern with this process is with the receipts themselves, as receipts aren't very structured and can vary from vendor to vendor. These points align with the characteristics of a cognitive automation opportunity also discussed in *Chapter 4*, *Identifying Cognitive Opportunities*:

- High variability

- Unstructured data

However, from a feasibility standpoint, we can easily create automation that handles receipts for the T&E analyst with UiPath Document Understanding. Also, from a viability standpoint, automating portions of this process for these analysts can provide the following benefits:

- **Increased throughput**: Implementing automation in the process can allow the analyst to complete reviews more quickly, reducing the 2-day turnaround time.

- **Increased capacity**: Implementing automation in the process can allow the analyst to complete more reviews, allowing the team to absorb the increased volume of expense reports without having to increase the team size or increase the turnaround time.

- **Error reduction**: Implementing automation can reduce errors made during manual review, minimizing the risk of approving an invalid expense.

Automation within this process can provide a great deal of impact for the T&E team. While there are many areas of opportunity, for this use case, we will focus on the receipt portion of the process. In the next section, we will focus on creating a future state design for implementing automation to process receipts.

Creating the future state design

When we look at the high-level design of the current state, there looks to be several automation opportunities that we can create future state automation for (*Figure 8.6*):

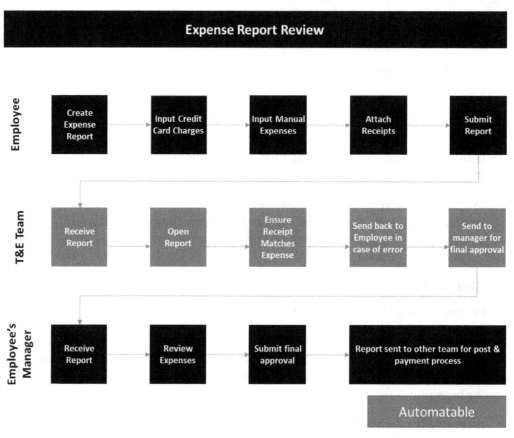

Figure 8.6 – Automation opportunities with the expense report review process

It appears that most of the steps taken by the T&E team in the process could potentially benefit from automation, through a mixture of automation ideas that hand off information to each other. However, since we want to focus on Document Understanding with this use case exercise, let's only focus on the steps of the process that appear between **Open Report** and **Send back to employee in case of error** (*Figure 8.7*):

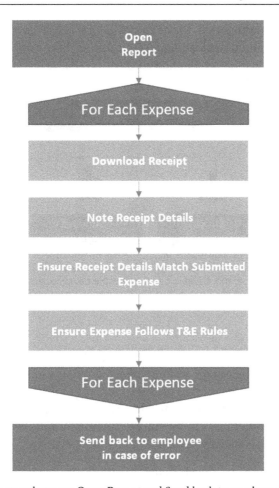

Figure 8.7 – The steps between Open Report and Send back to employee in case of error

As we can see, within the steps of **Open Report** and **Send back to employee in case of error**, we can apply Document Understanding, especially in the **Note Receipt Details** task of the process. So, we will focus on creating a future state design for this task.

In the future state of this automation, the analyst will be asked to continue manually, opening expense reports and downloading receipts as they have always done in the past; however, in the future state, they will launch automation that will read the contents of the receipts for them (*Figure 8.8*):

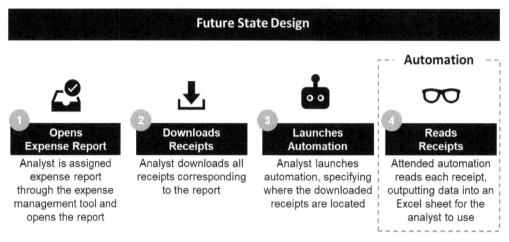

Figure 8.8 – The high-level future state design

For this use case, we are going to take an incremental approach to solving the pain points of the T&E team by using automation to read the contents of each receipt. By creating automation that can read receipts, we can tackle one of the tougher parts of the process, allowing us to expand to other tasks in the future, potentially automating the full end-to-end process through a series of automations.

Double-clicking into the read receipts task that we are automating, there is one large requirement to square away before creating the final future state – the fields we need to extract from each receipt. In the current state of the process, the analyst needs to note and confirm the following fields:

- Vendor Name
- Receipt Date
- Tax Amount
- Total Amount

These fields will become important as we build out the taxonomy of the use case, as we touched upon in *Chapter 6, Understanding Your Tools*.

To summarize – as the future state design of this use case, the project we intend to build will be an **attended automation** that prompts the end user for a folder location of downloaded receipts. The automation will then read and extract the relevant information from each receipt using UiPath Document Understanding, exporting the data into an Excel sheet on the end user's machine. The end user can then more quickly validate expense reports, as all receipts have been read and data has been gathered for the analyst.

Building the solution with the Document Understanding framework

As we venture into building our defined use case, we must ensure that we develop the automation according to best practice. As we build out the use case in the following sections, we will follow the steps of the UiPath *Document Understanding framework* (*Figure 8.9*):

Figure 8.9 – UiPath's Document Understanding framework

The steps of the Document Understanding framework allow us to best build cognitive automation, combining all the technologies and approaches necessary.

In addition to following the Document Understanding framework as we develop the use case, to help speed up development and to ensure we follow the framework, we will leverage UiPath's *Document Understanding Process template* during the development process. This template is a fully functional UiPath Studio project template based on a document processing flowchart for use on a simple demo or a large-scale implementation. The *Document Understanding Process template* leverages the components of the UiPath *Document Understanding framework*, coming preconfigured with a series of basic document types in taxonomy, a classifier configured to distinguish between these classes, and extractors to extract data. This best practice template can be adapted to easily kick off development.

Setting up the Document Understanding Process template

The *Document Understanding Process template* is available out of the box with UiPath Studio under *prebuilt templates*. While the template takes care of a lot of the heavy lifting for us, we still need to define a few settings.

Creating the project

To start a project with the Document Understanding Process template, complete the following steps:

1. Navigate to the **Templates** tab of UiPath Studio.
2. Select **Include Prerelease**.
3. Choose **Document Understanding Process**.
4. Click **Use Template**.
5. Name your project `Note Receipt Details`, and when prompted, provide a folder location where you would like to save the project.
6. Click **Create** to create the project and get started.

Choosing the **Document Understanding Process** template in UiPath Studio can be seen in *Figure 8.10*:

Figure 8.10 – Starting a project with the Document Understanding Process template

> **Important Note**
> The Document Understanding Process template is compatible with Studio versions 21.4.4 and greater.

Now that we have the project created for our use case, let's venture in to define the settings of the Document Understanding Process template.

Defining the settings of the template

Once the project has been created, there are a couple of settings we need to define to ensure that the automation runs to expectations.

The first settings we must define are to disable persistence support and background running. Since this use case is planned to be an attended use case, we will be presenting the Validation Station to the end user on their local machine. Thus, we won't need persistence support, since we won't be using UiPath Action Center to validate the results of Document Understanding. We also won't be needing background support, since the automation will be attended and, thus, run in the foreground.

The steps to disable persistence support and background running are as follows:

1. Navigate to the **Project** pane of the UiPath project and click **Project Settings** (*Figure 8.11*):

Figure 8.11 – Clicking Project Settings

2. Once **Project Settings** is open, click the toggle to the right of **Starts in Background** and **Supports Persistence** until disabled:

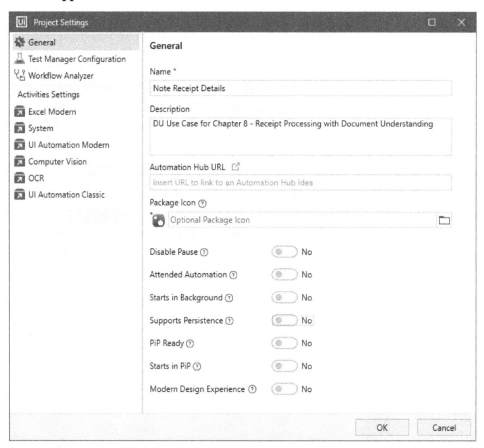

Figure 8.12 – Project Settings with persistence and background support disabled

Once persistence and background support are disabled, we need to configure the assets included with the *Document Understanding Process template* before we can get started building the use case. Assets are shared variables stored in UiPath Orchestrator that automation projects can reference as they execute. In the Document Understanding Process template, there are six assets that we need to assign values to in Orchestrator:

- **ApiKey**: Our Document Understanding API key.

- **AlwaysValidateClassification**: A boolean value telling UiPath whether a human will manually validate classification. If set to `True`, classification will always go through manual validation.

- **AlwaysValidateExtraction**: A boolean value telling UiPath whether a human will manually validate extraction. If set to `True`, extraction will always go through manual validation.

- **SkipClassifierTraining**: A boolean value telling UiPath to skip classifier training. If set to `True`, classifier training will not be performed.

- **SkipExtractorTraining**: A boolean value telling UiPath to skip extractor training. If set to `True`, extractor training will not be performed.

The steps to set the required assets are as follows:

1. Log into UiPath Orchestrator and navigate to a modern folder or personal workspace where the assets will be stored (*Figure 8.13*):

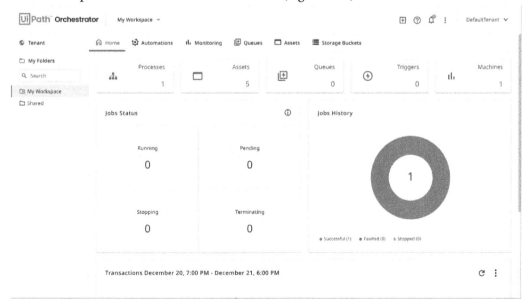

Figure 8.13 – The personal workspace in UiPath Orchestrator

2. Click on **Assets** to navigate to the **Assets** page (*Figure 8.14*):

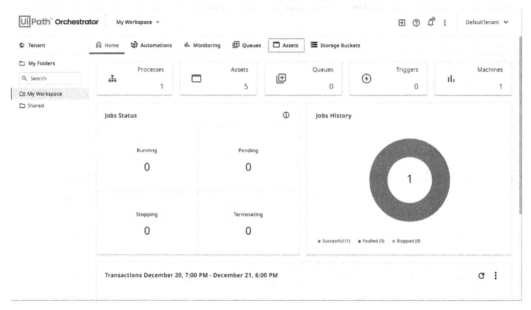

Figure 8.14 – Navigating to the Assets page

3. Add each asset individually by clicking the **Add asset** button; the settings should be those shown in *Figure 8.15*:

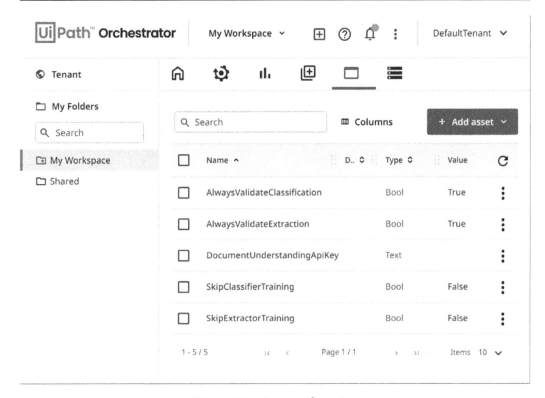

Figure 8.15 – Asset configuration

> **Important Note**
> Details for finding your API key for Document Understanding can be found
> at `https://docs.uipath.com/document-understanding/docs/api-key`.

Lastly, with our project settings set and assets defined, the last setting we need to define before venturing into building the use case is setting our main file. If you look at the project files of our newly created project, you'll notice two `.xaml` files:

- `Main-ActionCenter.xaml`
- `Main-Attended.xaml`

The reason for the two main files is that they each have small variances for unattended and attended executions. The `Main-ActionCenter.xaml` file contains Action Center activities that assist with validation for unattended scenarios. These actions aren't necessary in attended executions, since we can display the Validation Station to the user without the need for Action Center.

Since we are building an attended use case, we need to ensure that `Main-Attended.xaml` is set as the main file. Every UiPath project needs to have a defined main file, usually called `Main.xaml`, but in this case, we need to set `Main-Attended.xaml` as our main file by following these steps:

1. Right-click on **Main-Attended.xaml** underneath the project pane of UiPath Studio.

2. Click **Set as Main** to set the file as the project's main file (*Figure 8.16*):

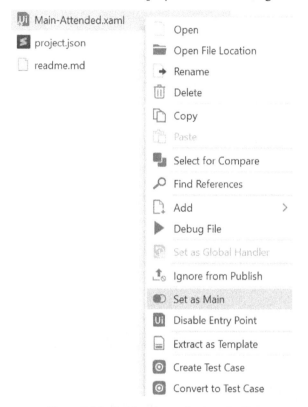

Figure 8.16 – Setting the project's main file

With the required settings of the *Document Understanding Process template* set, we can now move on to building the use case, with the first section of the *Document Understanding framework*, the **taxonomy**.

Creating the taxonomy

As the first step of the *Document Understanding framework*, we need to create a taxonomy to classify our documents and build a schema of the relevant information we want to extract from our receipts. As discussed in *Chapter 6, Understanding Your Tools*, a taxonomy is a system classification, and with Document Understanding, a taxonomy is a classification of **document types**.

From our current state understanding of the process, we defined the following as necessary for this process:

- Vendor name
- Receipt date
- Tax amount
- Total amount

Thus, as we build out our taxonomy, we must include these fields. One feature of the *Document Understanding Process template* is pre-built taxonomies for common documents, such as receipts. Let's review the out-of-the-box taxonomy and determine whether it fits our use case:

1. Open **Taxonomy Manager** within the design ribbon of UiPath Studio (*Figure 8.17*):

Figure 8.17 – UiPath Taxonomy Manager within the design ribbon

With Taxonomy Manager, note the pre-existing document types that are loaded – **Documents, Semi-StructuredDocuments**, and **Structured Documents**. Let's review the semi-structured documents.

2.   Click **Semi-StructuredDocuments** > **Financial** > **Receipt** (*Figure 8.18*):

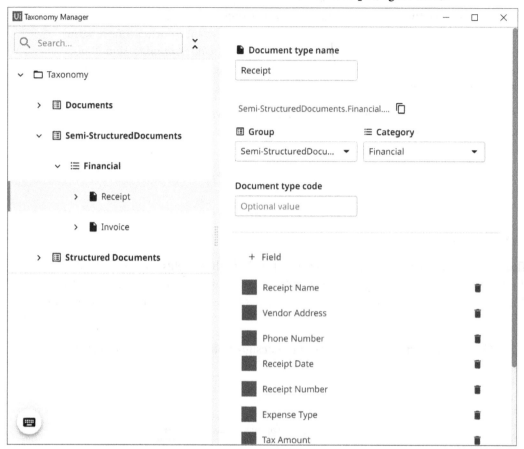

Figure 8.18 – The prebuilt Receipt document type

If we review the fields of the prebuilt **Receipt** document type, we can see that the fields match up with those from our current state understanding (*Figure 8.19*):

Taxonomy Design	
Fields from Current State understanding	**Fields from prebuilt Document Type**
Vendor Name	Receipt Name
Receipt Date	Receipt Date
Tax Amount	Tax Amount
Total Amount	Total Value

Figure 8.19 – Taxonomy Design

This mapping of the fields from our current state understanding against the fields of the prebuilt document type ensures that we don't have to edit the taxonomy at all. If there were additional fields that we wanted to capture, we could easily add the fields to the **Receipt** document type or create our very own document type. However, in this case, since we have a mapping, we can leverage the prebuilt document type and continue to the next section of the Document Understanding framework, **digitization**.

Setting up the digitizer

The next step of the *Document Understanding framework* is digitization. Digitization is the process of taking input data and scanning it into machine-readable text. Once the data is digitized, we can classify and extract the relevant data fields to create actionable steps for automation.

The steps of setting up digitization are mostly taken care of by the *Document Understanding Process template*. Within the project, open the 20_Digitize.xaml file to view the prebuilt digitization workflow, as shown in *Figure 8.20*:

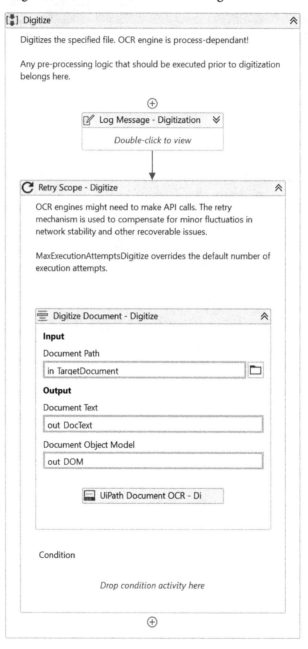

Figure 8.20 – The 20_Digitize.xaml workflow

Within the prebuilt workflow, we are given the opportunity to add any prework (such as adding gray scaling, resizing, or rotating) prior to digitizing the document. The prebuilt workflow then digitizes the document using the **UiPath Document OCR** as the OCR engine.

If we wanted to use a different OCR, we could simply replace the UiPath Document OCR with any of the OCR engines available. For this use case, we are going to use the UiPath Document OCR to digitize the receipts for us. Because of this, we don't need to add any additional activities for digitization, as the template does all the heavy lifting for us.

With the digitizer configured, we can now move on to the next section of the *Document Understanding framework*, which is **classification**.

Setting up the classifier

Once the digitizer is configured with our OCR engine of choice, we can move on to the next step of the *Document Understanding Framework*, classification. For this use case, we have only one document type (receipts), so classification may seem unnecessary, but let's set up UiPath Machine Learning Classifier to showcase how quickly it can be set up with the *Document Understanding Process template*.

When setting up classification, we must configure both the classifier itself and the trainer to retrain the classifier. Fortunately, as with digitization, the steps of classification are mostly taken care of by the Document Understanding Process template. To review the classification workflow of the template, open the `30_Classify.xaml` file to view the prebuilt classification workflow (*Figure 8.21*):

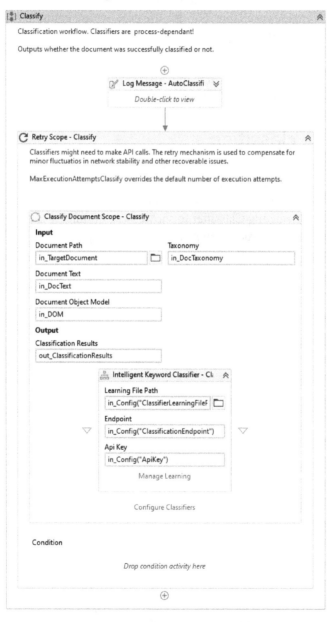

Figure 8.21 – The 30_Classify.xaml workflow

Within the prebuilt workflow, we are given the opportunity to add the relevant classifiers needed for a particular use case. By default, the Document Understanding Process template will include UiPath's Intelligent Keyword Classifier, but for this use case, let's replace the Intelligent Keyword Classifier with UiPath's Machine Learning Classifier.

The UiPath Machine Learning Classifier leverages an ML skill deployed to UiPath AI Center to classify a document into four classes – receipts, invoices, purchase orders, and utility bills. So, let's start with deploying the relevant ML skill to AI Center:

1. Navigate to **AI Center** within **Automation Cloud** and create a new project, Note Receipt Details.

2. Create a dataset, named DocumentUnderstanding_Classifier, to be leveraged by the ML skill for classifier training (*Figure 8.22*):

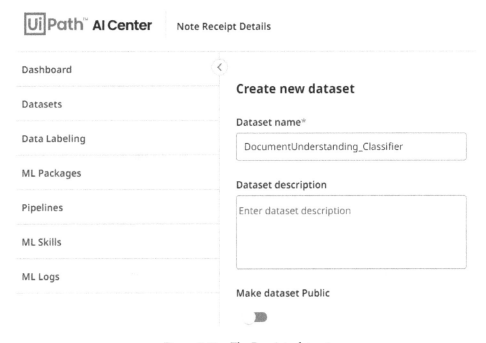

Figure 8.22 – The Receipts dataset

3. Deploy the **DocumentClassifier** out-of-the-box ML package, located within
    ML Packages > Out of the box Packages > UiPath **Document Understanding**
    (*Figure 8.23*):

> **Important Note**
>
> Choose the latest version of the **DocumentClassifier** skill. There's no need to
> include the optional fields, such as OCR.

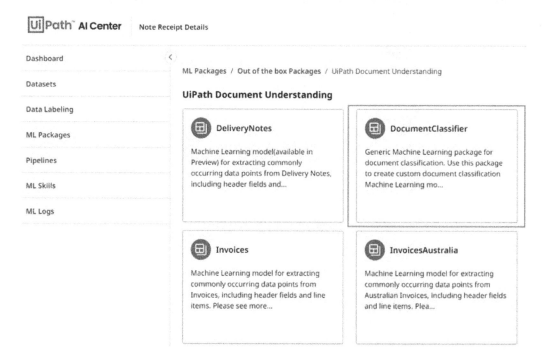

Figure 8.23 – The DocumentClassifier out-of-the-box package

With the dataset and the ML package created, we must next create a pipeline to be leveraged for continuously training our classifier. Navigate to **Pipelines** and create a new train run pipeline, using the dataset and ML package created earlier (*Figure 8.24*):

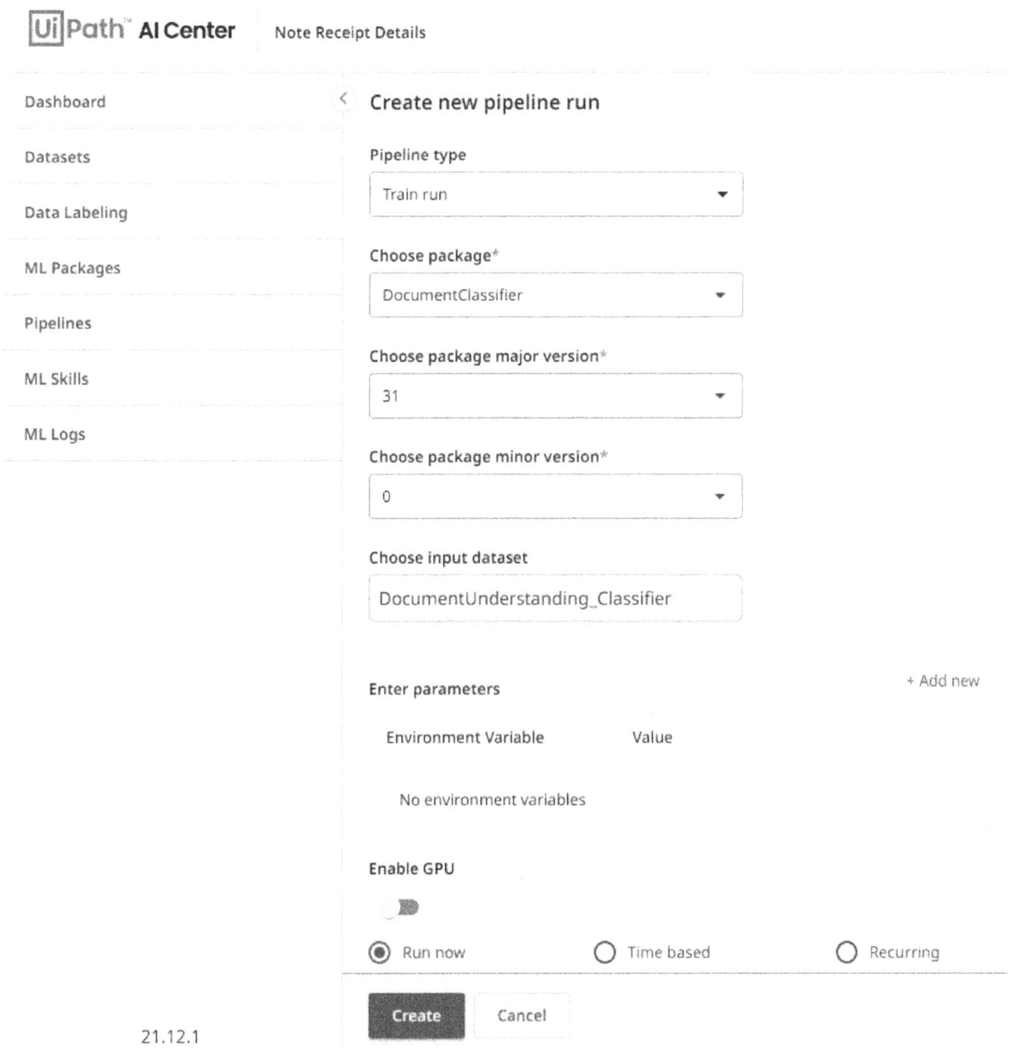

Figure 8.24 – Creating the training pipeline

With the training pipeline created, our last step is to deploy an ML skill to be leveraged in our UiPath project.

4. Navigate to **ML Skills** and create a new ML skill, using the ML package created earlier (*Figure 8.25*):

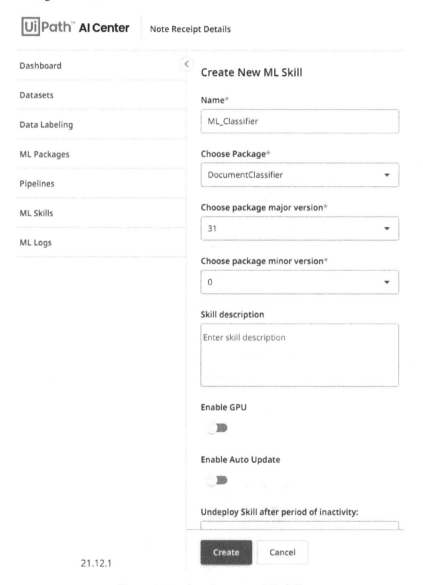

Figure 8.25 – Creating a new ML skill

Once the Machine Learning Classifier is deployed, we must next connect it to our UiPath project.

5. Navigate back to the workflow **30_Classify.xaml**.

6. Within **Classify Document Scope**, replace the **Intelligent Keyword Classifier** activity with a **Machine Learning Classifier** activity.

7. Within **Machine Learning Classifier**, choose the ML skill (**ML_Classifier**) that we just created (*Figure 8.26*):

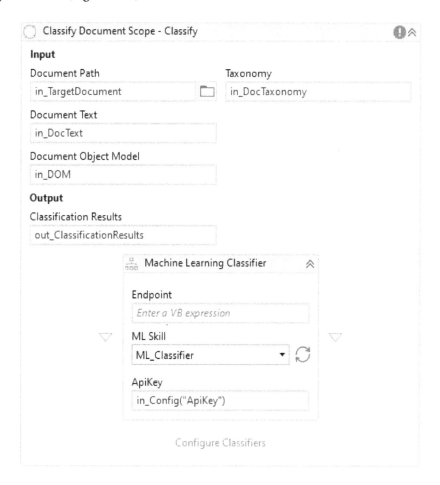

Figure 8.26 – Setting up the Machine Learning Classifier

Once the Machine Learning Classifier activity is added to the workflow, we must configure the classifier to classify the **Receipt** document type from our taxonomy.

8. Click **Configure Classifiers** and choose **receipts** for the **Receipt** document type (*Figure 8.27*):

> **Important Note**
> The Machine Learning Classifier is built for invoices, purchase orders, receipts, and utility bills, thus the dropdown contains the four choices.

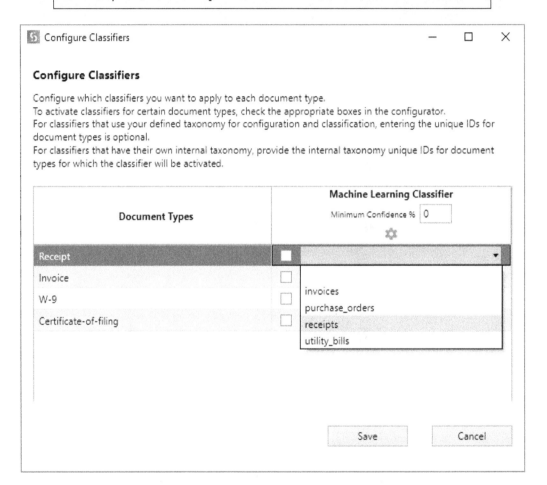

Figure 8.27 – Configure Classifiers

With the classifier deployed and connected to our UiPath project, the last step in classification is setting up the trainer. If we need to manually validate the classification from the UiPath robot, the trainer will retrain our ML skill with our manual input.

9. Open the **40_TrainClassifiers.xaml** workflow.

10. Within **Train Classifiers Scope**, replace the **Intelligent Keyword Classifier Trainer** activity with a **Machine Learning Classifier Trainer** activity.

11. Within the Machine Learning Classifier Trainer, choose the AI Center project (**Note Receipt Details**) and dataset (**DocumentUnderstanding_Classifier**) that we created earlier (*Figure 8.28*):

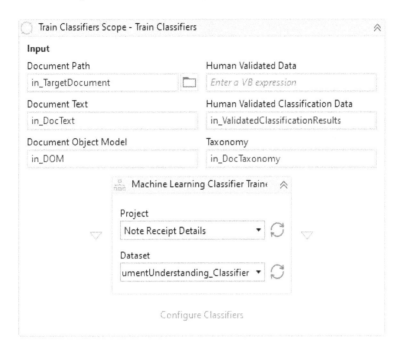

Figure 8.28 – Setting up the Machine Learning Classifier Trainer

Similar to the Machine Learning Classifier activity, after we add the Machine Learning Classifier Trainer, we must configure the classifier to classify the **Receipt** document type from our taxonomy.

12. Click **Configure Classifiers** and choose **receipts** for the **Receipt** document type (*Figure 8.29*):

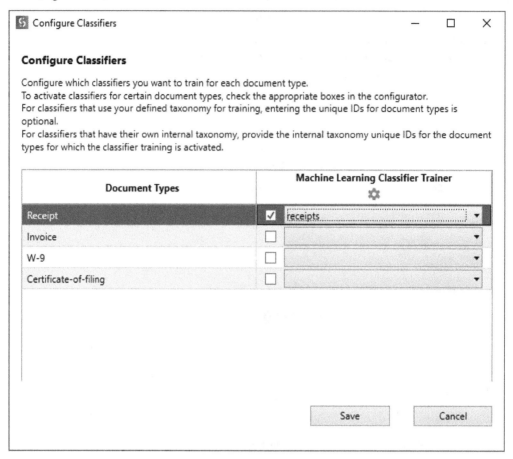

Figure 8.29 – Configure Classifiers

After configuring the classifier trainer, the configuration of the Machine Learning Classifier is complete.

We can now leverage an ML skill to classify receipts and continue to the next section of the Document Understanding framework, *extraction*.

Setting up the extractor

With classification completed, we can continue to the extractor portion of the Document Understanding framework. With extraction, automation will extract relevant data fields based on the taxonomy created at the beginning of the Document Understanding framework.

With the Document Understanding Process template, the extraction workflow is located within 50_Extract.xaml. Inside the **Data Extraction Scope** part of the workflow, note the preconfigured extractors included by the template (*Figure 8.30*):

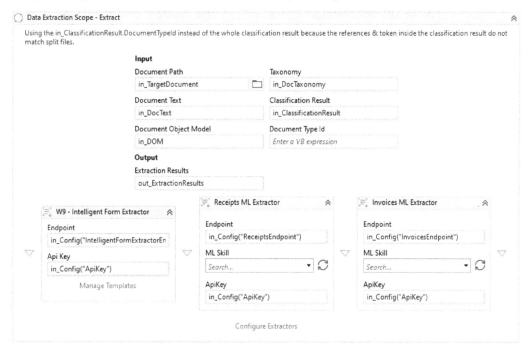

Figure 8.30 – Data Extraction Scope of 50_Extract.xaml

Unlike the Machine Learning Classifier, the Machine Learning Extractors for receipts and invoices can be hosted externally by UiPath or internally by ourselves via AI Center. Because of this, the Document Understanding Process template comes out of the box, already pointing to these externally hosted extractors. The UiPath Robot is instructed to use the extractor defined in the `ReceiptsEndpoint` key of the configuration file.

For this use case, we're going to host the **Receipts** ML Extractor on our AI Center so that we can retrain the extractor, as we did for classification. The steps to host the extractor are similar to the steps just performed for classification and are as follows:

1. Within the **Note Receipt Details** project in AI Center, create a new dataset, `Receipts`, to be leveraged for retraining.

2. Deploy the out-of-the-box ML package **Receipts**, located within **ML Packages** > **Out of the box Packages** > **UiPath Document Understanding** (*Figure 8.31*):

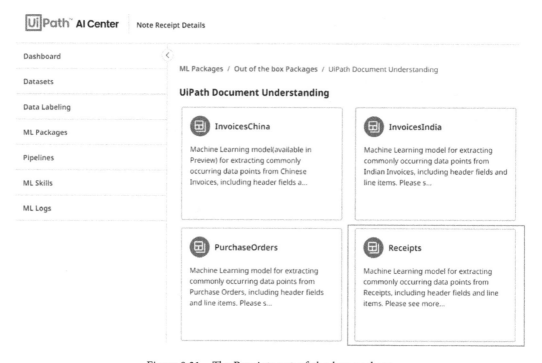

Figure 8.31 – The Receipts out-of-the-box package

3. Navigate to **Pipelines** and create a new train run pipeline, using the **Receipt** dataset and newly deployed **Receipt** ML package. Set the **auto_retraining** flag to `True` (*Figure 8.32*):

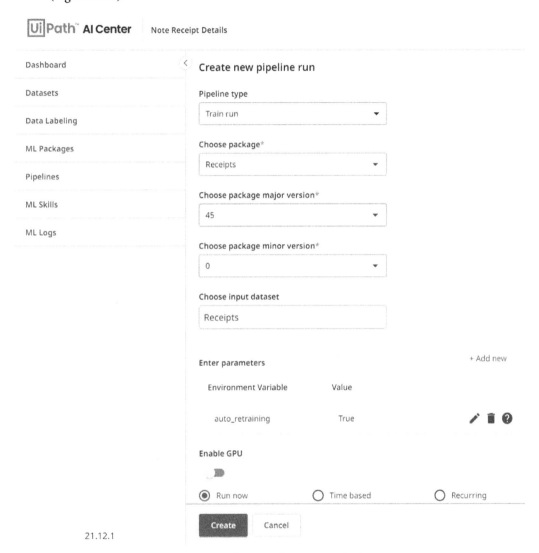

Figure 8.32 – Creating the training pipeline

With the training pipeline created, our last step is to deploy an ML skill to be leveraged in our UiPath project.

4. Navigate to **ML Skills** and create a new ML skill using the **Receipts** ML package created earlier.

    Once the ML Extractor is deployed, we must next connect it to our UiPath project.

5. Navigate back to the **50_Extract.xaml** workflow.

6. Within **Data Extraction Scope**, remove the endpoint within the **Receipts ML Extractor** activity and choose the **Receipts** skill underneath the **ML Skill** dropdown (*Figure 8.33*):

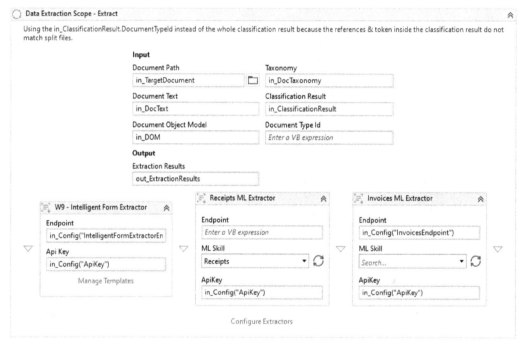

Figure 8.33 – The endpoint replaced with the deployed ML skill

With the Receipts extractor deployed and connected to our UiPath project, the last step in extraction is setting up the trainer.

7. Open the **60_TrainExtractors.xaml** workflow.

8. Within **Train Classifiers Scope**, select the **Note Receipt Details** project and **Receipts** dataset within the **Receipts Extractor Trainer** activity (*Figure 8.34*):

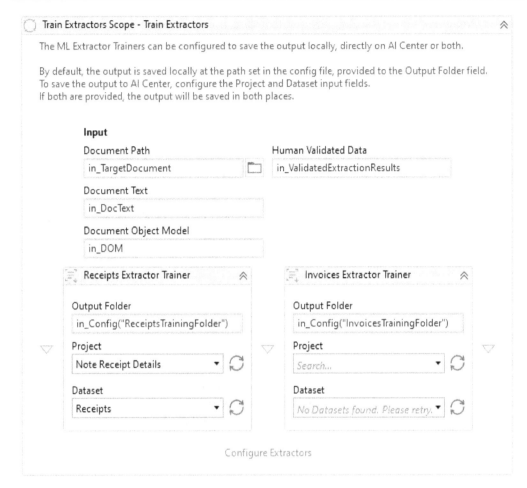

Figure 8.34 – Receipts Extractor Trainer

Because we are leveraging the existing ML Extractor Trainer activity for Receipts, the process template already handles configuring the extractor for us. Thus, the configuration of the ML Extractor is complete; we can now leverage an ML skill to extract information from classified receipts and can continue to the last section of the Document Understanding Framework, **exportation**.

Setting up the exporter

Once our data has been extracted, the last step of the Document Understanding framework is to export the extracted data into a medium that can be used by a human or another automation process. With exportation, automation will export relevant data fields based on the taxonomy created at the beginning of the Document Understanding framework.

With the Document Understanding Process template, the exportation workflow is located within `70_Export.xaml`. By default, this workflow will output an individual Excel sheet for each document processed through Document Understanding; however, for this use case, we want to export all our results into a single Excel sheet. To do this, we need to make a few modifications to the export workflow.

The first modification necessary is standardizing the filename of the exported Excel sheet. By default, the Document Understanding Process template will use a combination of the Receipt filename and page number, such as `1000-receipt_1-1.xlsx`. This works well for most use cases; however, we need to have one file to host the results of all the receipts:

1. Within the **Assign Output File Name** activity of `70_Export.xaml`, change the value to `Path.Combine(Environment.GetFolderPath(Environment.SpecialFolder.Desktop), "Note_Receipt_Details.xlsx")`.

    By changing the output filename to just `Note_Receipt_Details`, we have one file that we can export receipt data into. Changing the directory to the desktop will also directly save the file to the user's desktop folder.

    The second modification necessary is adding an **Append Range** activity to the workflow. Because the default Document Understanding Process template created an Excel sheet for each document, it only needed to use a **Write Range** activity to write a new Excel sheet for each document. If we wanted to store all results in one Excel sheet, using the **Write Range** activity would overwrite existing data, thus **Append Range** can be used to append data to the end of an existing Excel sheet.

2. Within **For Each Exported Table**, add an IF statement, checking whether **outputFile** exists by using the following condition – `File.Exists(outputPath)`.

3. Place the **Write Range** activity within the **Else** block of the **If** statement.

4. Drag an **Append Range** activity into the **Then** block of the **If** statement (*Figure 8.35*) with the following parameters:

    -   **DataTable**: `table`
    -   **SheetName**: `table.TableName`
    -   **Workbook Path**: `outputPath`

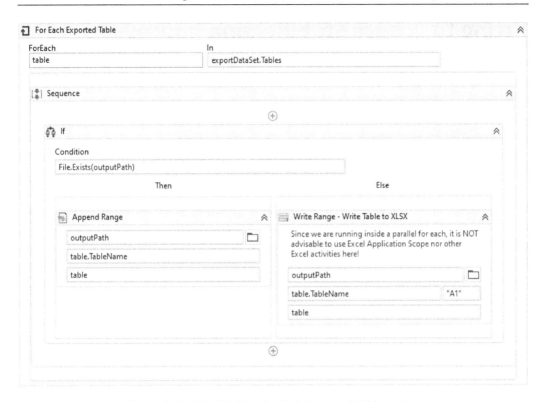

Figure 8.35 – Modified for the Each Exported Table activity

By adding the **Append Range** activity, the automation can add the results of a recently read receipt to the existing Excel workbook. If the workbook doesn't exist, the automation can revert to using the **Write Range** activity to create a new output file.

Do note that all exported data will be saved within `Data\Exports` of the project folder. This location can be changed by opening the `Config.xlsx` file located within the `Data` folder of the project and modifying the **ExportsFolder** entry.

After configuring the exporter, the configuration of the Document Understanding Process template is complete. Try running the workflow with the receipts at this link: `https://github.com/PacktPublishing/Democratizing-Artificial-Intelligence-with-UiPath/tree/main/Chapter08/Receipt%20Data`.

By following the steps of the *Document Understanding framework*, and using the *Document Understanding Process template*, we were able to quickly create an intelligent workflow to decipher receipts. In this section, we built the use case by leveraging the *Document Understanding Process template* and by following the *Document Understanding Framework* steps. In the next section, we will start testing our recently created workflow with sample data.

Testing to ensure stability and improve accuracy

With the initial development of the use case complete, we can venture into testing out how well the automation performs with the ML Classifier and ML Extractor. Testing any automated workflow before deployment is crucial in order to ensure that the automation works as expected. In this section, we will investigate enabling the Validation Station for the ML Classifier and ML Extractor, as well as starting testing with sample data.

Enabling the Validation Station

During the development of the use case earlier, we deployed both the **DocumentUnderstanding** classifier and the **Receipts** ML skill to act as our classifier and extractor respectively. One of the reasons why we deployed these skills to AI Center was the ability to retrain these skills using the Validation Station. This gives us the ability to manually validate automation performance and retrain ML models, something we call **Closing the Feedback Loop** (*Figure 8.36*):

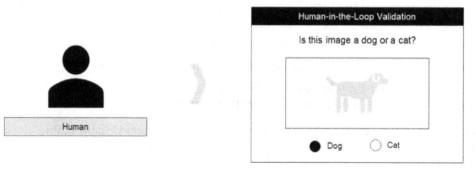

Figure 8.36 – Closing the Feedback Loop

As discussed in *Chapter 7, Testing and Refining Development Efforts*, using the Validation Station to close the feedback loop can play a large factor in improving the performance of an ML model.

By deploying both the **DocumentUnderstanding** classifier and **Receipts** extractor to AI Center, and by creating datasets within AI Center to store validation data, we can easily enable the Validation Station to close the feedback loop on our use case.

When we were defining the settings of the Document Understanding Process template, we configured a couple of assets in our UiPath Orchestrator, specifically the following:

- **AlwaysValidateClassification**: A boolean value telling UiPath to always go through manual validation for classification. If set to True, validation will always be performed.

- **AlwaysValidateExtraction**: A boolean value telling UiPath to always go through manual validation for extraction. If set to True, validation will always be performed.

During the configuration of the Document Understanding Process template settings, we set both of those assets to True (*Figure 8.37*), meaning that we were telling UiPath that we did not want to skip classifier or extractor training; instead, we wanted UiPath to launch the Validation Station and take our human input to retrain the classifier and extractor:

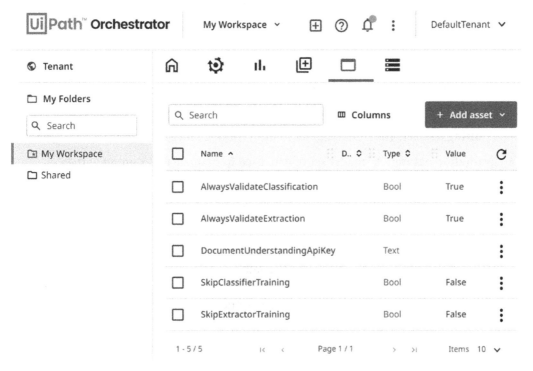

Figure 8.37 – Asset configuration

For testing purposes, the Validation Station should be enabled to help retrain the extractor and classifier, thus both assets should be set to `True`. Once testing is completed, you may want to turn off validation by setting the assets to `False`, or by adding additional logic to the workflow to only present the Validation Station when the classifier or extractor returns low confidence. Another recommendation is to work with the business and agree on a success metric – for example, testing 100 scenarios, resulting in a 90% positivity rate, before the Validation Station is disabled.

Testing with sample receipts

With the Validation Station enabled, we can now test the performance of the deployed classifier and extractor and retrain both skills where necessary. To start testing, navigate to the following link to retrieve sample receipts: `https://github.com/ PacktPublishing/Democratizing-Artificial-Intelligence-with- UiPath/tree/main/Chapter08/Receipt%20Data`.

At the preceding link, you will find receipts to test with and receipts for use after testing is complete. We will be leveraging the receipts located within the `Testing` folder for the purpose of testing.

In *Chapter 7, Testing and Refining Development Efforts*, we discussed that when training cognitive automation we should have multiple datasets to test our ML skill:

- **Training dataset**: The initial dataset to train the ML model
- **Evaluation dataset**: A smaller set of data used to evaluate the training of the ML model
- **Testing dataset**: An unbiased dataset used to expose the trained ML model to real-life data

Fortunately, for this use case, the ML skills we used (the **DocumentUnderstanding** Classifier and the **Receipts** Extractor) already come pre-trained and ready to use. That means that UiPath has already performed the initial training and evaluation for us, leaving us to only test the dataset with our unbiased data, retraining with the Validation Station where necessary.

Within the testing folder, you will find 50 receipts to test with. Feel free to test with a subset of the 50 receipts, or to individually test with all 50. When testing, be sure to validate the classification (*Figure 8.38*) and extraction (*Figure 8.39*) to further strengthen the ML skills deployed to AI Center:

Figure 8.38 – Validating classification

When using the Validation Station for classification, be sure to choose **Receipt** in the dropdown. You should then notice the percentage change to 100%; that is because we manually validated the classification and are 100% certain that the document is of the receipt document type. Once validated, you can click **Save** to continue, and UiPath will use that validation to retrain the **DocumentUnderstanding** classifier.

> **Important Note**
>
> Further information on how to interact with Classification Station can be found at https://docs.uipath.com/activities/docs/present-classification-station#using-classification-station.

As with classification, you should manually validate all fields returned from extraction (*Figure 8.39*):

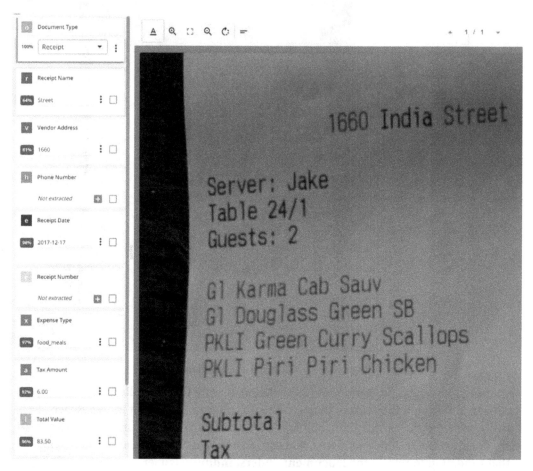

Figure 8.39 – Validating extraction

Once validated, you should then notice the percentage change to 100%. There may be some instances where the extractor did not correctly extract the correct information, such as **Receipt Name** in *Figure 8.39*. Note that the extractor only extracted **Street** instead of **1660 India Street**. This is a tough receipt, since the receipt name looks like a vendor's address.

To correct this extraction, hover over the text with your mouse and press *Ctrl + left-click*. That will place a red dotted rectangle around the text you want to validate (*Figure 8.40*):

Figure 8.40 – Validating Receipt Name

Then, click on the three dots under the **Receipt Name** field and click **Change extracted value** to change the **Receipt Name** value to our validated value (*Figure 8.41*):

Figure 8.41 – Receipt Name validated

Depending on the structure of the receipt and potentially on its image, you may need to change some of the extracted fields. For other receipts, you may not need to change any values. If you don't need to change any values, you should still validate the extracted value by checking the checkbox, changing the percent confidence to **100%**. This is true for missing values and table items as well, as all validation helps retrain and strengthen the model.

> **Important Note**
>
> Further information on how to interact with the Validation Station can be found at `https://docs.uipath.com/activities/docs/` `present-validation-station#using-the-validation-` `station`.

Once validated, you can click **Save** to continue, and UiPath will use that validation to retrain the **DocumentUnderstanding** classifier. Continue testing the automation with additional receipts from the testing folder. As you continue testing and retraining, performance should increase with the classifier and extractor.

In summary, using the Validation Station to validate classification and extraction results is crucial during the testing phase. Any human validation is an opportunity to make our ML skills smarter and stronger. In this section, we learned how to enable the Validation Station within the *Document Understanding Process template* and then learned how to interact with the Validation Station. In the next section, we will prepare the use case for deployment by adding features to enhance the end user experience.

Deploying with the end user experience in mind

With the initial development and testing of the use case complete, we can venture into deploying the automation into a production environment. However, before deploying into production, there are a few items we should add to the automation in order to improve the end user experience. In this section, we will create a **dispatcher**, an additional workflow to dispatch our `Main-Attended.xaml` file. After creating the dispatcher, we will add a couple of dialog boxes to keep the end user informed about the status of the automation. Once complete, we will be ready for the last section, deploying the automation into production.

Creating the dispatcher

You may have noticed from testing the use case during the last section that we can only run the automation for one receipt at a time. Because the *Document Understanding Process template* is configured to handle one document at a time, we need to create a mechanism that dispatches the Document Understanding Process template. The dispatcher will ask the user for a folder path as input and, for each file within the folder, call the Document Understanding Process template we created, meeting the requirements set with the future state design.

To create the dispatcher, we must first create a new workflow:

1. In the **Note Receipt Details** project, create a new workflow file called `Dispatcher`.

2. Once created, set the dispatcher file as the main one (*Figure 8.42*):

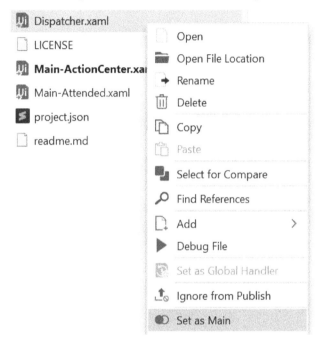

Figure 8.42 – Setting Dispatcher.xaml as the main file

Once the dispatcher is created and set as the main file of the project, we can start adding code to the workflow.

3. Within `Dispatcher.xaml`, drag a **Browse for Folder** activity into the main sequence, saving the output as a variable called `selected_folder`.

4. Next, drag a **For Each File in Folder** activity into the sequence, setting the **In Folder** input property to `selected_folder`.

5. Within the **Do** sequence of **For Each File in Folder**, drag an **Invoke Workflow File** activity, setting the **WorkflowFileName** input property as `Main-Attended.xaml`, the file we created earlier (*Figure 8.43*).

6. Within the **Invoke Workflow File** activity, click **Import Arguments** and set the
    `in_TargetFile` parameter to `CurrentFile.FullName`. This will ensure that
    the Main-Attended workflow runs with each file within the folder:

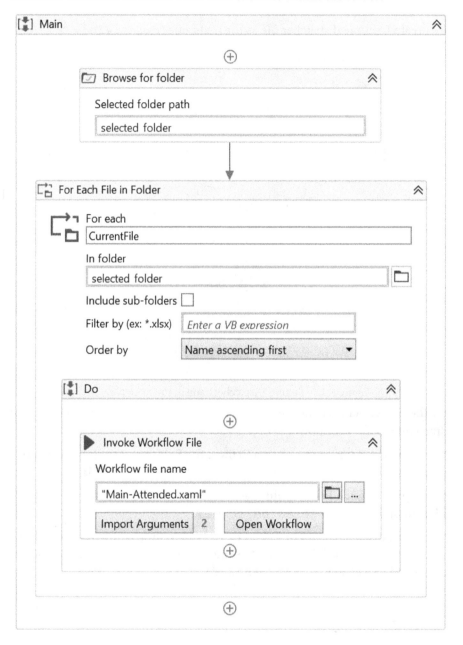

Figure 8.43 – The completed Dispatcher.xaml workflow

By adding an additional workflow with a **For Each File in Folder** activity, we can continuously call the `Main-Attended.xaml` workflow we created with the *Document Understanding Process template*. In this section, we created a new workflow, **Dispatcher**, and set the file as the main file of the project. Within `Dispatcher.xaml`, we added a folder picker activity to prompt the end user for a folder path. We then added a for-each loop to loop through the files of the folder, calling `Main-Attended.xaml` for each receipt. In the next section, we will add a few additional prompts to guide the end user through the automation execution before we deploy the automation into production.

Adding user input prompts

Best-practice attended automation should always contain prompts or message boxes to guide and/or inform the user on the progress of automation. As discussed in *Chapter 5, Designing Automation with End User Considerations*, making automation easy to work with is an easy way to ensure automation adoption. In this section, we will work on adding message boxes and information dialogs to keep the end user informed during automation execution.

When creating the dispatcher in the last section, we added a folder picker to prompt the human user to choose a folder of receipts for automation to extract. A new user may find this folder picker confusing, as they may not know what to choose, so we should add a message box before the folder picker, guiding the user on what folder to pick:

1. Within `Dispatcher.xaml`, drag a **Message Box** activity before the **Browse for folder** activity (*Figure 8.44*).

2. Add a descriptive message inside the **Text** input property, such as `Please choose a receipts folder`:

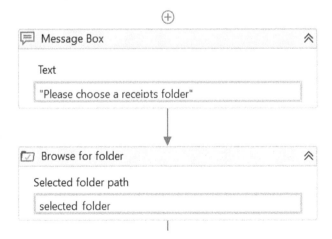

Figure 8.44 – Adding a message box to the Dispatcher

By adding a message box prior to presenting the folder picker, we can more accurately guide the user in choosing the correct folder for automation to interact with. While adding a message box is one way to keep the user informed, another method is to present status bars, keeping the user updated on the status of automation (*Figure 8.45*):

Figure 8.45 – Presenting a status bar

The UiPath team has created a set of custom activities exactly for the purpose of keeping the user updated, through status bars, progress bars, and message boxes. With the **Interactive Activities** custom activity, we can quickly and easily add these messages to our workflow.

> **Important Note**
>
> More information on the Interactive Activities can be found at `https://marketplace.uipath.com/listings/notification-activities`.

To install the Interactive Activities in your project, follow these steps:

1. Click **Manage Packages** in the **Design** ribbon of the **Note Receipt Details** project.
2. Within the **Manage Packages** window, click **Marketplace** and search for `UiPathTeam.Interactive.Activities` (*Figure 8.46*):

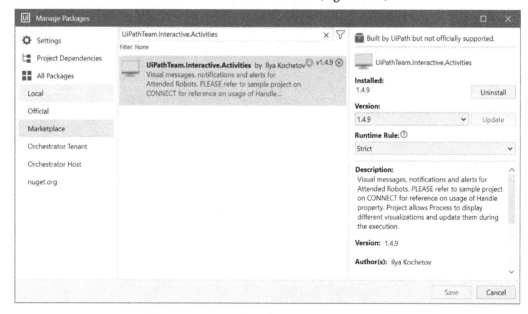

Figure 8.46 – Installing the Interactive Activities

3. To install, choose the latest version and click **Save**.

   Once the Interactive Activities are installed, they can be added to the project. To best show the user's progress, we should present the status bar when starting and completing a receipt, as well as when going through the milestones of the Document Understanding framework (digitization, classification, extraction, and exportation). Let's start by adding a starting message to the dispatcher.

4. Drag a **Display Message** activity inside the **Do** sequence of the **For Each File in Folder** activity.

5. Within the **Display Message** activity, add a descriptive message within the **Message** input property, and create a `windowIdentifier` variable within the **Handle** output property.

   The **Handle** output property is very important to include if you want to replace the status bar message multiple times. The handle acts as a unique identifier to the status bar window, and not setting the property while making updates to the status bar will result in multiple status bars appearing on top of each other (*Figure 8.47*):

Figure 8.47 – Status bars stacked on top of each other

6. Within the `Variables` pane, set the `windowIdentifier` scope to the main sequence instead of **Do** (*Figure 8.48*). This ensures that the unique identifier is not reset with each iteration of the for loop:

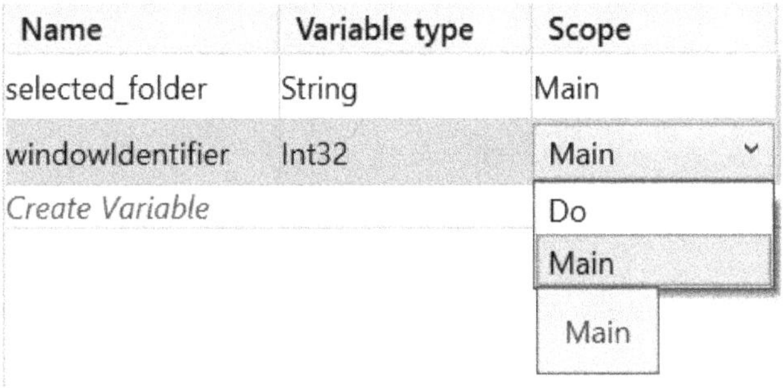

Name	Variable type	Scope	
selected_folder	String	Main	
windowIdentifier	Int32	Main	˅
Create Variable		Do	
		Main	
		Main	

Figure 8.48 – Setting the scope of windowIdentifier

Once the **Display Message** activity is added, your dispatcher workflow should look like *Figure 8.49*:

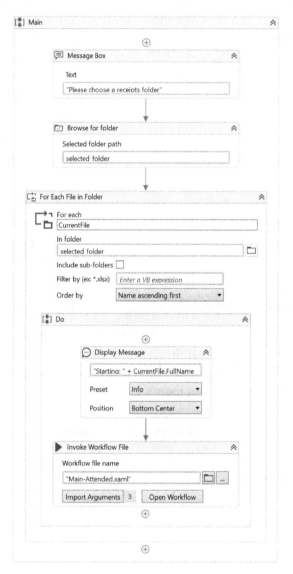

Figure 8.49 – The dispatcher workflow with a display message

Additional status messages can be added inside the Document Understanding Process template to call out digitization, classification, extraction, and exportation. You can add **Display Message** activities to each of those individual workflows; just ensure to pass along windowIdentifier set in the dispatcher workflow to keep statuses from stacking on top of each other.

By adding message boxes and status messages, we have made the automation a little more end user-friendly by giving them guidance on what folder to choose and keeping them informed on the status of the automation. With these activities now added to the workflow, we can call the automation complete and move on to the next section of deploying the automation into production.

Deploying into production

In a production setting, we shouldn't run our automation directly from UiPath Studio, nor should we expect our end users to have access to UiPath Studio. Thus, we must deploy our automation into UiPath Orchestrator to allow others to run our automation from their UiPath Assistant.

The UiPath Assistant is a desktop tool for users to launch and interact with UiPath automations, a *launchpad* of sorts (*Figure 8.50*):

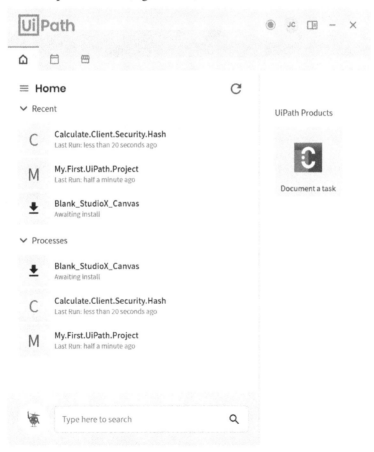

Figure 8.50 – The UiPath Assistant

When automation is deployed for mass use, it is deployed into UiPath Orchestrator, and then it will appear on end users' assistants for them to run.

> **Important Note**
>
> More information on deploying automation can be found at `https://docs.uipath.com/studio/docs/about-publishing-automation-projects`.

For this use case, we're going to quickly publish the automation into our own personal workspace within our UiPath Assistant:

1. Within the **Note Receipt Details** project, click the **Publish** button in the **Design** ribbon of UiPath Studio.

2. Check that the **Publish to** field is set to **Orchestrator Personal Workspace Feed** (*Figure 8.51*):

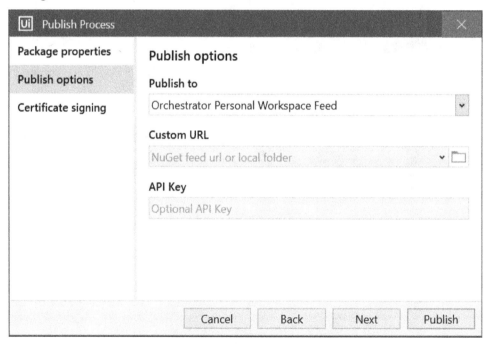

Figure 8.51 – Publishing the automation

3. Click **Publish**.

Once published, you should be able to see the automation within your UiPath Assistant (*Figure 8.52*):

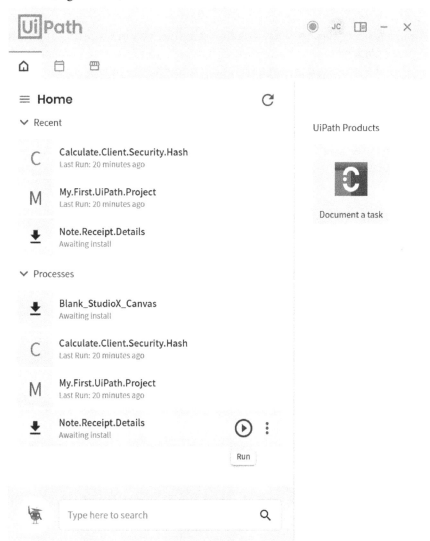

Figure 8.52 – The project deployed to the UiPath Assistant

Clicking the **Run** button to the right of **Note.Receipt.Details** will install the automation on your machine and start execution.

Once deployed, you are free to run the automation directly from the UiPath Assistant. Any changes to the code can be made in UiPath Studio and then republished by following the steps just discussed. In this section, we completed the use case build by deploying the **Note Receipt Details** project into our UiPath Assistant – great job!

Summary

In this chapter, we covered building a cognitive automation use case with *UiPath Document Understanding*. In the construction of the use case, we leveraged the *UiPath Document Understanding Process template* to build in line with best practice following the *UiPath Document Understanding framework*. By having a full understanding of Document Understanding and the Document Understanding framework, we built a cognitive use case that can decipher receipts for the T&E team to use during their expense report review process.

In the next chapter, we will take another hands-on approach by working together to build the second of three use cases with UiPath. We will learn how to use UiPath's AI Center to classify emails.

9
Use Case 2 – Email Classification with AI Center

A classic use for **machine learning (ML)** is text classification. Every organization, irrespective of the industry, has a use case for classifying text. Examples of text classification use cases include routing emails to corresponding folders or routing service requests to relevant teams. A lot of this work is still manual, requiring an individual to review the text and use their judgment to interpret and classify the text's contents. Fortunately, with UiPath's AI Center, we can create cognitive automation to apply intelligent understanding of text, potentially reducing the need for manual work.

By having a full understanding of AI Center, we will work together to build cognitive automation that can interpret and classify text. As we build a use case, we will follow a similar development life cycle as outlined in the previous chapters. We will start with understanding the current state of our automation opportunity, then design the future state of the process with automation involved. Once the future state is designed, we will build automation with UiPath Studio and AI Center, test the solution to ensure stability, and finally deploy the automation into production.

In this chapter, *Use Case 2 – Email Classification with AI Center*, we will cover the following topics as we build automation to process receipts:

- Understanding the current state
- Creating a future state design
- Building a solution with AI Center
- Testing to ensure stability and improve accuracy
- Deploying with the end-user experience in mind

Technical requirements

All code examples for this chapter can be found on GitHub at the following link:

```
https://github.com/PacktPublishing/Democratizing-Artificial-
Intelligence-with-UiPath/tree/main/Chapter09
```

Working with UiPath is very easy, and with the Community version, we can get started for free. However, for UiPath AI Center, we will need an Enterprise license. You can acquire a 60-day Enterprise trial license from uipath.com.

With the 60-day Enterprise trial, you will have access to the following:

- Five RPA Developer Pro licenses—named user licenses include access to Studio, StudioX, Attended Robot, Apps, Action Center, Task Capture
- Five Unattended Robots, five Testing Robots, five AI Robots
- AI Center, AI Computer Vision, Automation Hub, Data Service, Document Understanding, and Insights

For this chapter, you will require the following:

- UiPath Enterprise Cloud (with AI Center)
- UiPath Studio 2021.4+
- The UiPath.MLServices.Activities package v1.2.0 or higher
- The UiPath.WebAPI.Activities package v1.9.2 or higher

> **Important Note**
>
> Directions on how to install UiPath packages can be found at the following link:
>
> `https://docs.uipath.com/studio/docs/managing-activities-packages`

Check out the following video to see the Code In Action at: `https://bit.ly/3udG6qL`

Enabling AI Center in UiPath Enterprise trial

UiPath Document Understanding comes out of the box with the UiPath Community and Enterprise versions; however, with UiPath AI Center, we need to enable the service within the Enterprise trial. You can enable the service by following the steps here:

1. Navigate to Automation Cloud at `https://cloud.uipath.com`. The Automation Cloud home page looks like this:

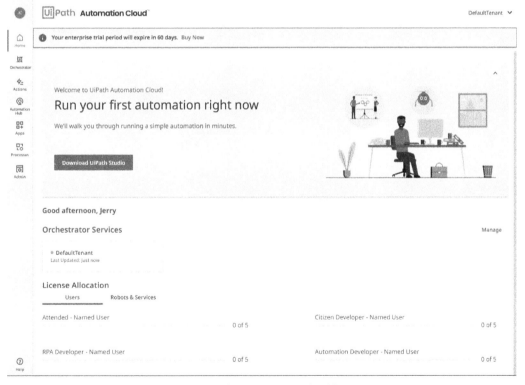

Figure 9.1 – UiPath Automation Cloud home page

2. Navigate to **Admin**, as highlighted in the following screenshot:

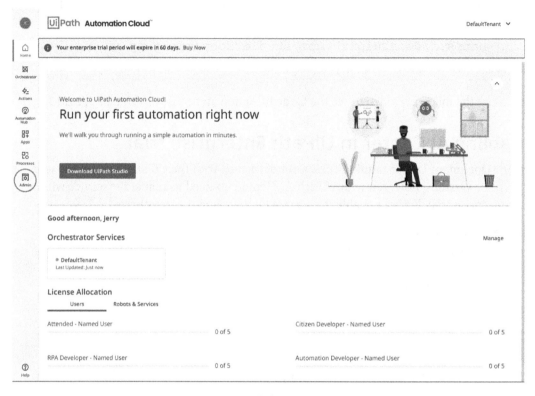

Figure 9.2 – Clicking on Admin

3. In **DefaultTenant**, click on **Tenant Settings**, as highlighted in the following screenshot:

Figure 9.3 – Clicking on Tenant Settings

4. Choose **AI Center** under **Provision Services**, and click on **Save**. The process is illustrated in the following screenshot:

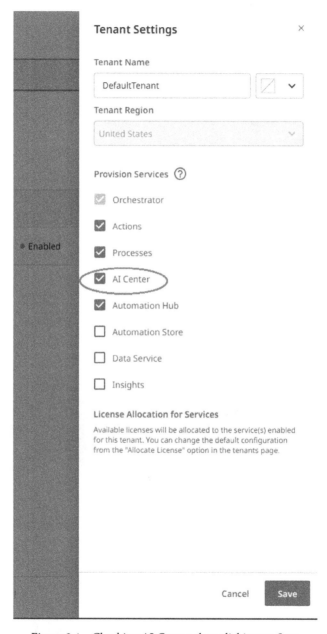

Figure 9.4 – Checking AI Center, then clicking on Save

Once AI Center is enabled, we have completed the technical requirements for this chapter. In the next section, we will start gathering context about the use case by reviewing the current state.

Understanding the current state

Let's start this use case by learning a little about the current state of the process. Every day, the accounting team of *Company ABC* receives invoices submitted by vendors as part of its **procure-to-pay** (**P2P**) process, as illustrated in the following screenshot:

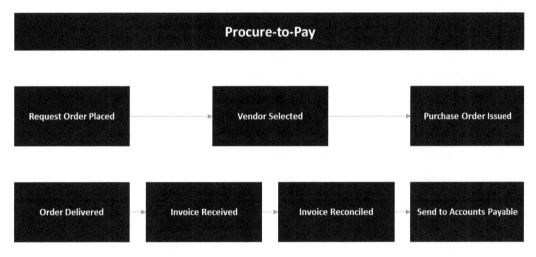

Figure 9.5 – Company ABC's P2P process

As defined by *Company ABC*, vendors must send invoices to a central inbox that the accounting team maintains. As the accounting team receives these invoices, they are reviewed for correctness and entered into their system of record, where they are reconciled and forwarded to **Accounts Payable** (**AP**) for payment, if all is correct.

Company ABC has been doing extremely well this year, thus has been purchasing more goods and services than in years past. This influx in purchasing has put a strain on the accounting team as it tries to maintain an efficient P2P process. A portion of this strain has been caused by the increase of spam emails received within the shared email inbox for invoices. The team wants to be able to effectively filter out spam emails so that it can focus on invoices.

Looking at the current state, there appears to be plenty of opportunity for automation—reverting to the characteristics of an automation opportunity from *Chapter 4, Identifying Cognitive Opportunities*, we can see that this automation opportunity has the following characteristics:

- **Manual interactions with applications**: Accounting analysts manually interact with the shared email inbox.

- **Repetitive in nature**: The same steps are performed for each email received in the shared inbox.

- **Time-consuming**: Reviewing spam emails and cleaning the inbox can be time-consuming.

From a feasibility standpoint, we can easily create automation that monitors an inbox for incoming emails, classifying between emails that are spam and not spam. Also, from a viability standpoint, automating portions of this process for these analysts can provide the following benefits:

- **Increased capacity**: Implementing automation can allow analysts to spend more time performing high-value tasks, rather than cleaning an email inbox.

- **Error reduction**: Implementing automation can create a cleaner inbox, ensuring that an invoice email is not missed by the accounting team.

Automation within this process can provide a great deal of impact for the accounting team. Creating automation to filter out spam within an email inbox is one of the many potential use cases we can build for an accounting team. In the next section, we will focus on creating a future state design for implementing automation to process receipts.

Creating a future state design

With our new understanding of the current state and the pain points of the accounting team, we can start with creating a future state design of the automation. Since this automation requires no human interaction, creating a future state design for this use case is generally straightforward, as depicted in the following screenshot:

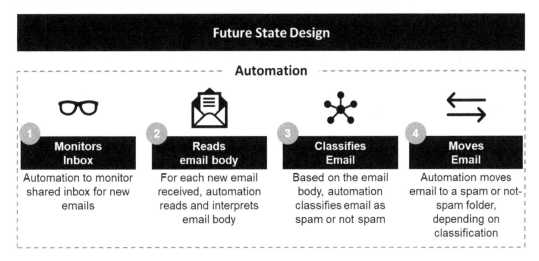

Figure 9.6 – High-level future state design

For this use case, automation will have the responsibility of monitoring the accounting team's shared inbox. The automation can run in the background in an attended or unattended scenario, scanning the shared inbox for new emails. Once an email is received, the automation will retrieve the email, read its contents, and classify the email as spam or not spam based on the contents of the email body. Lastly, based on the classification, the automation will route the email to a corresponding folder where an accounting analyst can review it. The folders that automation will route to are listed here:

- `Invoices`
- `Low Confidence`
- `Spam`
- `Retrain-Not Spam`
- `Retrain-Spam`

The reason for the `Low Confidence` email folder is to have a landing zone for any email that automation has low confidence in classifying. We wouldn't want to have a scenario where an invoice is accidentally placed in the `Spam` folder, so any email the automation has difficulty with will be routed to the `Low Confidence` folder. We'll also be adding a retraining component to the use case, where incorrectly classified emails can be moved to one of the `Retrain` folders to retrain our ML skill.

To summarize, as the future state design of this use case, the project we intend to build can be leveraged in either an attended- or unattended-type scenario. The automation will monitor an email inbox and route incoming emails based on their body contents. Based on the automation's classification, incoming emails will be routed to one of three folders. The end user can then more quickly review invoices, without needing to continually clean the shared inbox with this automation.

Building a solution with AI Center

As we venture into building a use case, we can split development into the following two sections:

1. Creating an ML skill
2. Building automation workflows

We will start with deploying our ML skill into AI Center, and then create an automation workflow in UiPath Studio. Let's begin with creating an ML skill.

Creating an ML skill

In this section, we will create and train an ML skill for our use case with AI Center. By deploying an ML package and by loading a dataset to train and evaluate our ML model, we will build an ML skill ready to be tested with our automation workflow.

Creating an AI Center project

Before we can deploy an ML package or create an ML skill, we must first create a project in AI Center. To do so, follow these steps:

1. Navigate to UiPath Automation Cloud and click on **AI Center** to get started, as highlighted in the following screenshot:

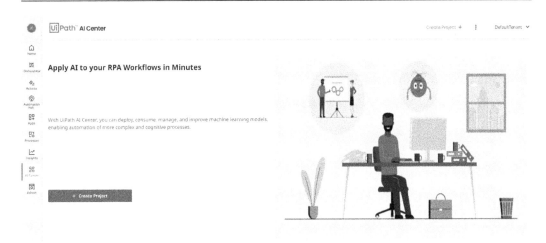

Figure 9.7 – AI Center home page in UiPath Automation Cloud

2. Once navigated to AI Center, click **+ Create Project** and provide a descriptive project name and description.

Once your project is created, we can start with loading our datasets to train and evaluate the ML model.

Loading datasets

In AI Center, a dataset is merely a collection of files and folders creating a set of data to be leveraged by an AI Center project. Within **Datasets**, we can upload any set of files and folders necessary to allow our ML models to access data.

For this use case, we're going to create two separate datasets for our training and evaluation pipelines. The dataset we'll be using is the *SMS Spam Collection* dataset from the **University of California Irvine** (**UCI**) *Machine Learning Repository*, which can be found at the following link: https://archive.ics.uci.edu/ml/datasets/ SMS+Spam+Collection.

While **Short Message Service (SMS)** text is a little different than email text, it should still prove out the use case of filtering out spam versus not spam. For this project, we have taken the *SMS Spam Collection* dataset and split the data into *Training, Evaluation, Testing,* and *Production* datasets, which can be found at the following GitHub link: `https://github.com/PacktPublishing/Democratizing-Artificial-Intelligence-with-UiPath/tree/main/Chapter09/Email%20Data`.

> **Important Note**
> The dataset has been split into small and large datasets. The large dataset is the full set, while the small dataset is just a subset of the large dataset. Feel free to use either, but note that training with the large dataset will take a lot longer than with the small dataset.

Before uploading our data to AI Center, let's take a quick glance here at the training dataset:

text	label
Go until jurong point, crazy.. Available o	ham
Ok lar... Joking wif u oni...	ham
Free entry in 2 a wkly comp to win FA Cu	spam
U dun say so early hor... U c already the	ham
Nah I don't think he goes to usf, he lives	ham
FreeMsg Hey there darling it's been 3 we	spam
Even my brother is not like to speak wit	ham
As per your request 'Melle Melle (Oru M	ham
WINNER!! As a valued network custome	spam

Figure 9.8 – Training dataset

With the dataset, we have two columns, **text** and **label**, which contain the following information:

- `text` is the plain text of the SMS message.
- `label` is the classification of either `spam` or `ham`.

Now that we have a little bit of context on our datasets, let's add our training dataset to AI Center by following the next steps:

1. Within the project dashboard, click **Datasets** to navigate to the **Datasets** page.
2. Click **Create New** to create a new dataset, providing a name of `Email Classification` for our dataset.

3. After the dataset is created, click **Upload** to upload the training and evaluation folders from the following GitHub link to AI Center: `https://github.com/ PacktPublishing/Democratizing-Artificial-Intelligence- with-UiPath/tree/main/Chapter09/Email%20Data`.

Once uploaded, the `Email Classification` dataset should look like this:

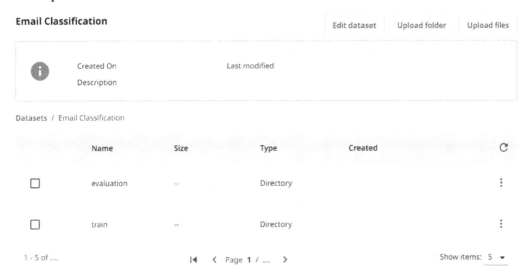

Figure 9.9 – Email Classification dataset

With our datasets loaded into AI Center, we can now proceed with deploying an ML package and running our pipelines.

Creating an ML package

As mentioned in *Chapter 6, Understanding Your Tools*, an ML package in AI Center is a folder containing all the code and associated files necessary to serve an ML model.

Because we're performing a text classification exercise, we're going to use UiPath's out-of-the-box `EnglishTextClassification` ML package. Here are the steps you need to follow:

1. Navigate to **ML Packages** by clicking the **ML Packages** tab in the project dashboard.

2. When presented with **Create a new package**, click **Out of the box Packages**.

3. Underneath **Out of the box Packages**, choose **Language Analysis** > **EnglishTextClassification**, choosing **Package Version 6.0**.

After providing the ML package with a name, you should see your new ML package in your project dashboard. With our package created, let's investigate creating a pipeline that can train our newly deployed ML package.

Creating pipelines

An ML pipeline is a series of steps taken to produce an ML model. During the pipeline process, we will provide an input, and once completed, the pipeline will produce a set of useable outputs and logs that lead to the creation of an ML skill.

In AI Center, there are three types of pipelines, as outlined here:

- **Training pipeline**: Produces a newly trained **ML Package** version from an input ML package and dataset

- **Evaluation pipeline**: Produces a set of metrics and logs from a newly trained **ML Package** version and dataset

- **Full pipeline**: A full data process pipeline, performing both training and evaluation pipelines

For this use case, let's use our newly deployed training and evaluation datasets to run a training and evaluation pipeline. Proceed as follows:

1. Navigate to **Pipelines** by clicking the **Pipelines** tab in the project dashboard.

2. Click **Create New** to create a new pipeline, selecting the following:

   - **Pipeline type**: `Train run`

   - **Choose package**: `TextClassification`

   - **Choose package major version**: `6`

   - **Choose package minor version**: `0`

   - **Choose input dataset**: `Email Classification/train/`

   - **Enable GPU**: Turn this off

   Under **Environment Variable**, add the following environment variables:

   - `input_column`: `text`

   - `target_column`: `label`

The following screenshot provides a representation of this:

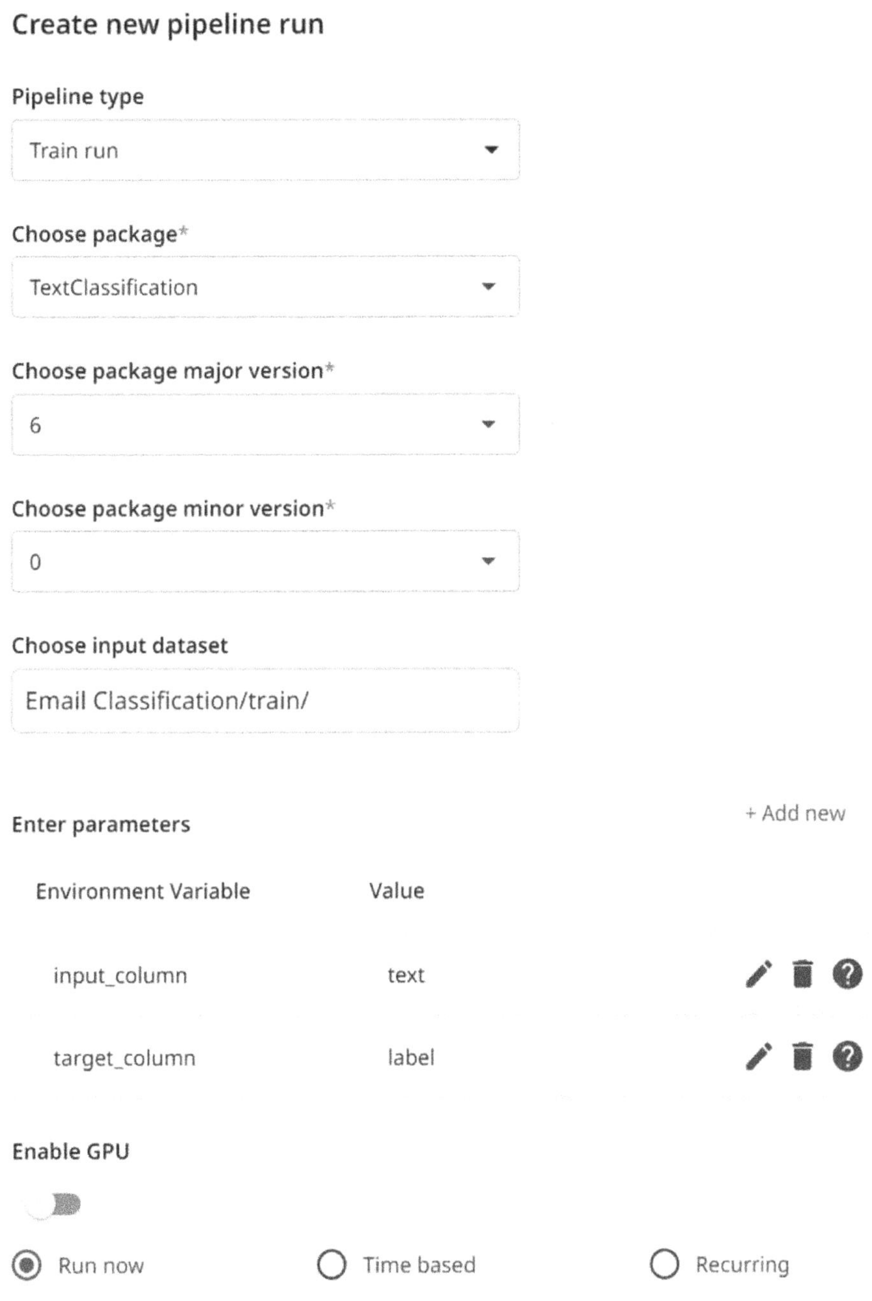

Create new pipeline run

Pipeline type

Train run ▼

Choose package*

TextClassification ▼

Choose package major version*

6 ▼

Choose package minor version*

0 ▼

Choose input dataset

Email Classification/train/

Enter parameters   + Add new

Environment Variable	Value
input_column	text
target_column	label

Enable GPU

◉ Run now   ◯ Time based   ◯ Recurring

Figure 9.10 – Creating a training pipeline

3. Click **Create** to start the training pipeline.

> **Important Note**
>
> The size of the dataset will affect the pipeline duration time. Choosing **Enable GPU** can lead to quicker output, but **graphics processing units (GPUs)** are not available with UiPath Community Edition.

4. Perform *Steps 2*, *3*, and *4* again, choosing Evaluation run as the pipeline type, version 6.1 of the TextClassification package, and the evaluation dataset folder as the input dataset. Your screen should now look like this:

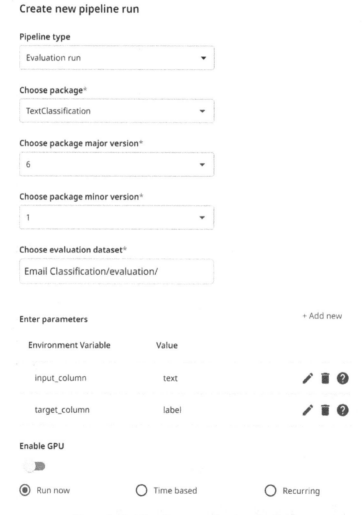

Figure 9.11 – Creating an evaluation pipeline

Once completed, the evaluation pipeline will return a few artifacts as output, as follows:

- `Evualation.csv`: Data used to evaluate the model
- `Evaluation-report.pdf`: Output report of the pipeline run

If you look at the evaluation report, you can see that pretty good performance can be achieved with relatively little data.

The **key performance indicator** (**KPI**) to look for in the evaluation report is to achieve a greater-than-80% Fscore for binary classification problems or a greater-than-70% F-score for multi-class problems. In this case, we just reached 80%; however, if we wanted to increase this number, we could increase the size of our dataset or close the feedback loop with a retraining mechanism, which we'll build later.

And that's it! In the next section, we'll deploy the ML skill.

Deploying the ML skill

Once deployed into AI Center, an ML skill can be consumed by automation. Choosing the **ML Skills** activity in UiPath Studio, you can select from a list of ML skills deployed on AI Fabric, passing input data and leveraging output data as needed.

With our model trained and evaluated, we can now deploy it into a skill usable with **robotic process automation** (**RPA**). Proceed as follows:

1. Navigate to **ML Skills** by clicking the **ML Skills** tab in the project dashboard.

2. Click **Create New** to create a new ML skill, selecting the following:

   - **Name**: `Email_Spam_Classification`
   - **Choose Package**: `TextClassification`
   - **Choose package major version**: 6
   - **Choose package minor version**: 1
   - **Enable Auto Update**: Turn this on

Your screen should now look like this:

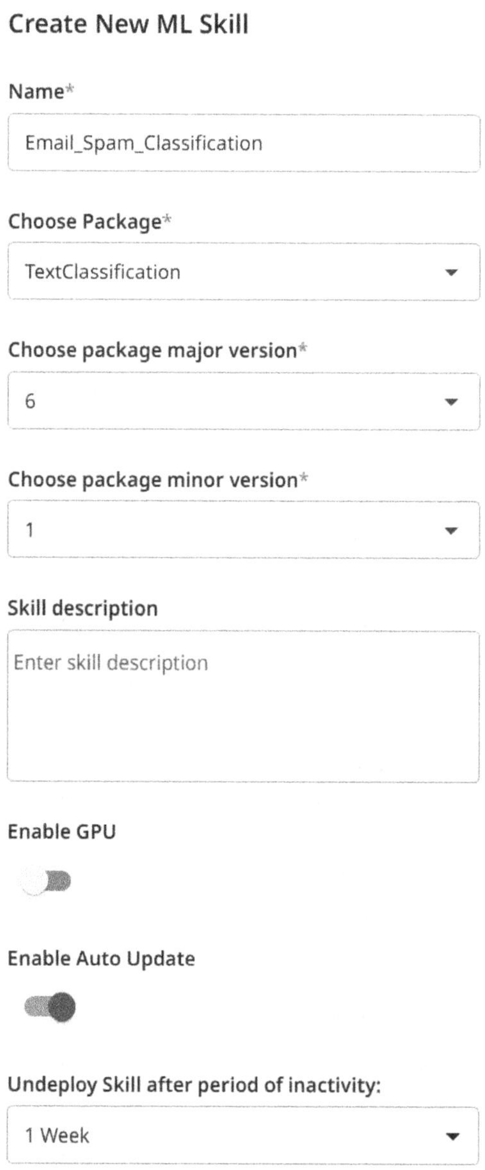

Create New ML Skill

Name*

Email_Spam_Classification

Choose Package*

TextClassification   ▼

Choose package major version*

6   ▼

Choose package minor version*

1   ▼

Skill description

Enter skill description

Enable GPU

Enable Auto Update

Undeploy Skill after period of inactivity:

1 Week   ▼

Figure 9.12 – Creating an ML skill

3.   Click **Create** to create an ML skill.

With our ML skill deployed in AI Center, we can move on to connect it with automation by building out our automation workflows. By way of review, in this section, we created a project in AI Center, loaded our datasets, created an ML package, created a training and evaluation pipeline, and—lastly—deployed an ML skill.

Building automation workflows

With our ML skill deployed, we can move our focus on to creating automation workflows for the use case. Approaching this use case, we will build our project in a few steps, as follows:

1. Creating a project
2. Building an ML workflow
3. Building supplemental workflows
4. Building the **Main** workflow

After completing these steps, we will have the foundation of our project complete and ready for testing.

Creating a project

For this project, we'll create a flowchart workflow using the **Transactional Process** template (*Figure 9.13*) in Studio. To start a project with the **Transactional Process** template, complete the following steps:

1. Navigate to the **Templates** tab of UiPath Studio.
2. Choose **Transactional Process**.
3. Name your project Email Classification, and provide a folder location to save the project to.
4. Click **Create** to create a project and to get started.

The **Transactional Process** template is illustrated in the following screenshot:

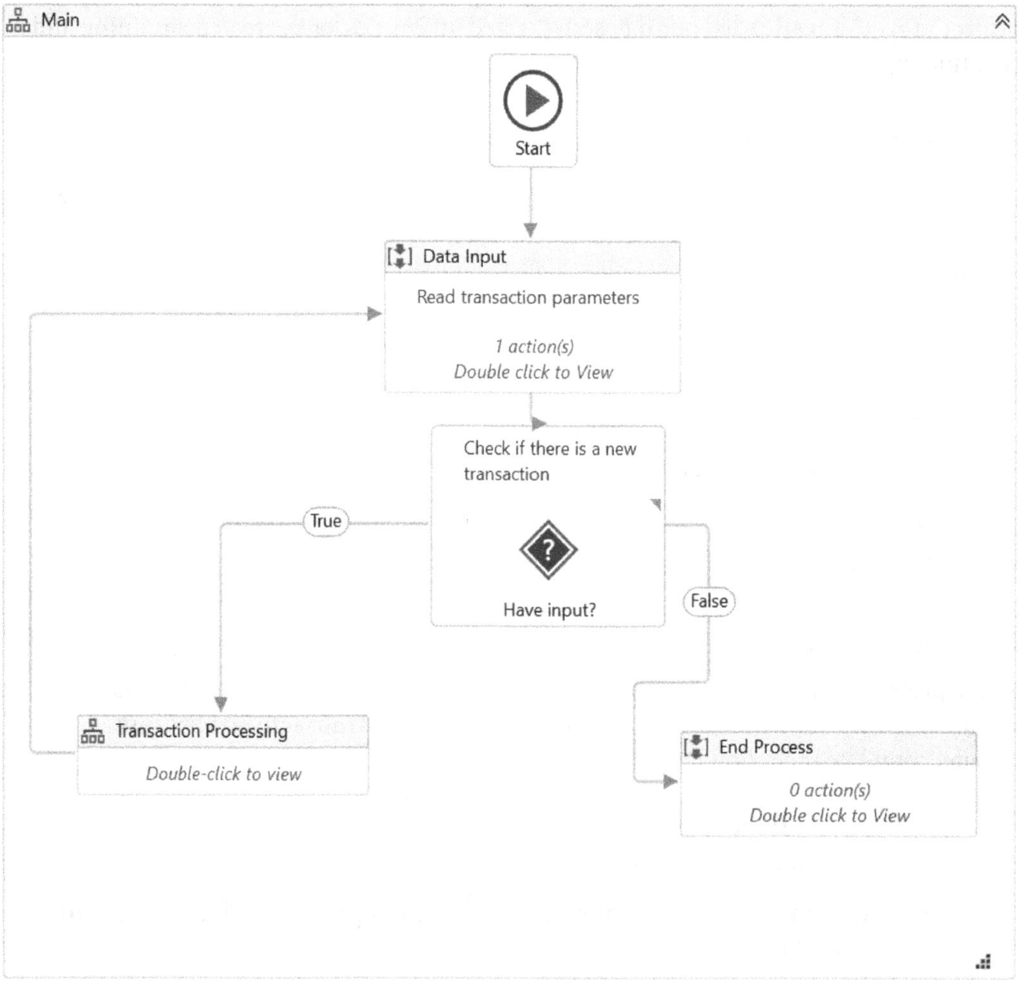

Figure 9.13 – Transactional Process template

As you open the Main.xaml file of our newly created project, you will be presented with the **Transactional Process** template, a flowchart designed for basic automation projects. Within the **Main** flowchart, you will notice several premade containers, as outlined here:

- **Data Input**: The purpose here is to retrieve the data necessary for running the project. In our use case, we'll use this sequence to retrieve emails.

- **Have input?**: The purpose here is to check if input data exists. In this use case, we'll use this Flow Decision to check if any new emails were retrieved.

- **Transaction Processing**: The purpose here is to perform a set of actions for each transaction. In this use case, we'll use this flowchart to read and classify our email.

- **End Process**: The purpose here is to perform any closing activities. In this use case, we'll start by leaving it blank, but we may use it for testing later.

Once our project is created, click **Manage Packages** in UiPath Studio and add the following dependencies into the project:

- `UiPath.MLServices.Activities`

- `UiPath.WebAPI.Activities`

That concludes creating a project. We'll revert to the **Main** workflow in a later section where we build out the **Main** workflow, but for now, we'll transition into building ML and supplemental workflows.

Building an ML workflow

With the automation project created, we can focus on creating some supplemental workflows that'll become building blocks of our **Main** workflow. In the ML workflow, our objective is to create a modular workflow sequence that takes an email body as input, passes it through our newly deployed ML skill, and returns the classification and confidence of the skill.

To start, let's create a new workflow, as follows:

1. In the `Email Classification` project, create a new workflow file called `Classify_Email`.

2. Once the new workflow is created, create three arguments, as illustrated in the following screenshot:

Name	Direction	Argument type	Default value
in_EmailText	In	String	*Enter a VB expression*
out_Classification	Out	String	*Default value not supported*
out_Confidence	Out	Double	*Default value not supported*
Create Argument			

| Variables   Arguments   Imports | | | 🖐 🔎 100%   ▾ 🔲 🔳 |

Figure 9.14 – Classify_Email arguments

The purpose of each argument is this:

- `in_EmailText`: To input text from our email's body
- `out_Classification`: To output a classification text of `Spam` or `Ham`
- `out_Confidence`: To output a classification double as a percentage of confidence of the classification

Once the arguments are added to the workflow, we can start adding activities to the blank canvas.

3. Add a **Log Message** activity, with `in_EmailText` as the message and `Info` as the log level. This will allow the automation to log the text of the email body to the console when we run the automation—potentially helpful if we need to debug anything down the road.

4. After the **Log Message** activity, add an **ML Skill** activity with the following parameters:

- **Connection Mode**: `Robot`
- **ML Skill**: `Email_Spam_Classification`
- **Item**: `in_EmailText`
- **JSON Response**: `json_response`

Within the **ML Skill** activity, `json_response` should be a newly created variable of type `String`. This variable is responsible for holding the **JavaScript Object Notation (JSON)** response of our `Email_Spam_Classification` ML skill.

With `json_response` being a string of key-value pairs, before we can query the JSON, we need to convert it into an object through deserialization.

5. Drag a **Deserialize JSON Response** activity into the workflow with the following parameters:

- **Json String**: `json_response`
- **Type Argument**: `System.Collections.Generic.Dictionary<System.String, System.Object>`
- **Json Object**: `json_object`

Similar to `json_response`, `json_object` should be a newly created variable of type `System.Collections.Generic.Dictionary<System.String, System.Object>`. This will allow us to now query our JSON response with a set of keys.

The JSON response of `Email_Spam_Classification` is going to return a prediction and a confidence that we need to assign to `out_Classification` and `out_Confidence`, respectively.

6. Drag an **Assign** activity into the workflow, with the following parameters:

    - **To**: `out_Classification`
    - **Value**: `json_object("prediction").ToString`

7. Drag another **Assign** activity into the workflow, with the following parameters:

    - **To**: `out_Confidence`
    - **Value**: `Convert.ToDouble(json_object("confidence"))`

    With our last **Assign** activity dragged into the workflow, the `Classify_Email` activity is almost complete. To help with potential debugging in the future, let's add one more **Log Message** activity to log our classification and confidence from the ML skill.

8. Add a **Log Message** activity, with `out_Classification` and `out_Confidence` within the message, and `Info` as the log level.

With our final log message added, the `Classify_Email` workflow is complete. It should look similar to this:

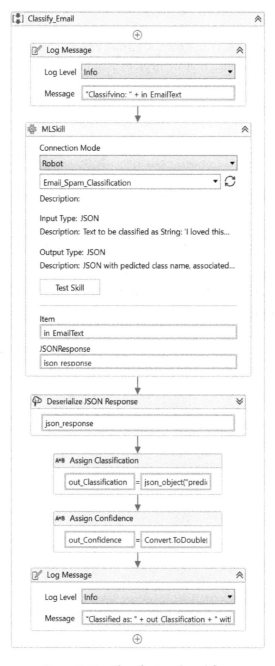

Figure 9.15 – Classify_Email workflow

If you want to test the workflow, set the default value of in_EmailText to a set of random text, as illustrated in the following screenshot, and try running the workflow:

Name	Direction	Argument type	Default value
in_EmailText	In	String	"hello, world!"
out_Classification	Out	String	*Default value not supported*
out_Confidence	Out	Double	*Default value not supported*
Create Argument			

Figure 9.16 – Setting the default value

As you run the workflow, take notice of the **Output** pane of UiPath Studio and the log messages we added. It should look like this:

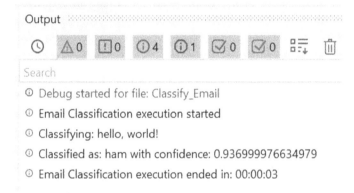

Figure 9.17 – Output of hello, world! test

That concludes our development of the ML workflow. In this section, we created a workflow that accepts a String email body as an input, passes that input through our ML skill, deserializes the JSON response, and outputs the skill's confidence and classification. In the next section, we'll work on some of the other supplemental workflows needed for our project before we piece it all together in our **Main** workflow.

Building supplemental workflows

Before we move on to putting the whole project together in the **Main** workflow, there's an additional supplemental workflow that we can build. The purpose of this workflow is to accept an email message and email folder as input and move that email message into the designated email folder.

The reason for creating this workflow separately, outside of our **Main** workflow, is based on the thought that code that's copied and pasted is best off in its own workflow versus being pasted multiple times into another workflow. By having a separate workflow for moving emails, within the **Main** workflow we can simply invoke this supplemental workflow whenever we need to move an email into separate folders.

To start, let's create a new workflow, as follows:

1. In the `Email Classification` project, create a new workflow file called `Move_Email`.

2. Once the new workflow is created, create two arguments, as illustrated in the following screenshot:

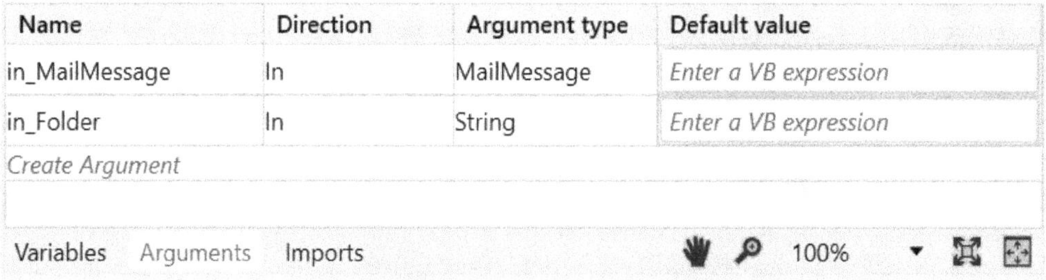

Name	Direction	Argument type	Default value
in_MailMessage	In	MailMessage	*Enter a VB expression*
in_Folder	In	String	*Enter a VB expression*
Create Argument			

Variables   Arguments   Imports   ✋ 🔎 100%   ▾ 🔲 🔳

Figure 9.18 – Move_Email arguments

The purpose of each argument is this:

- `in_MailMessage`: To input the mail message from our current email
- `in_Folder`: To input the folder location for our destination folder

Once the arguments are added to the workflow, we can start adding activities to the blank canvas.

3. Add a **Log Message** activity, with `in_Folder` as the message and `Info` as the log level. This will allow the automation to log the destination folder to the console when we run the automation—potentially helpful if we need to debug anything down the road.

4. Drag a **Move Mail Message** activity into the workflow, with the following parameters:

- **MailFolder**: `in_Folder`
- **MailMessage**: `in_MailMessage`

5. Add a final **Log Message** activity with `Info` as the log level, and a message stating success.

    After adding the final **Log Message** activity, the `Move_Email` workflow should look like this:

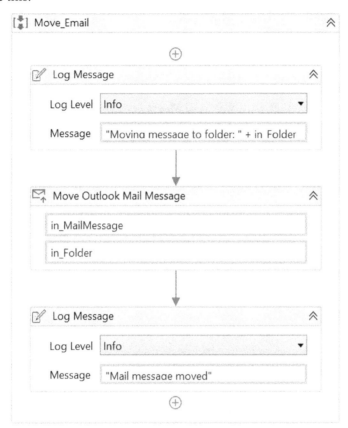

Figure 9.19 – Move_Email workflow

While this workflow may seem simplistic in nature, it'll become very helpful as we build the **Main** workflow and need to move emails based on our classification and confidence. In this section, we created a supplemental workflow that moves an email message based on the desired email folder. In the next section, we'll work on putting these building blocks together in our **Main** workflow.

Building the Main workflow

Every automation project starts with a **Main** workflow. The purpose of the **Main** workflow is to kick off the project, invoking other workflows in the process.

From our future state design, the first step of the use case is retrieving emails from our inbox—we can recreate this by adding a folder to Outlook, named `Email Classification`, and having automation fetch emails from that folder. Proceed as follows:

1. Open the **Data Input** sequence.

2. After **Assign**, drag a **Get Outlook Mail Messages** activity into the workflow, setting the following attributes:

   - **MailFolder**: `Inbox/Email Classification`
   - **MarkAsRead**
   - **OnlyUnreadMessages**
   - **Top**: `1`
   - **Output**: `list_emails`

3. Drag an **If** statement into the workflow, setting the condition to `List_emails.count > 0`.

4. Within the **Then** sequence of the **If** statement, drag an **Assign** activity into the workflow.

5. Within the **Assign** activity, set **NewTransaction** to `True`.

After adding the **Assign** activity within the **If** statement, our **Data Input** sequence should look similar to this:

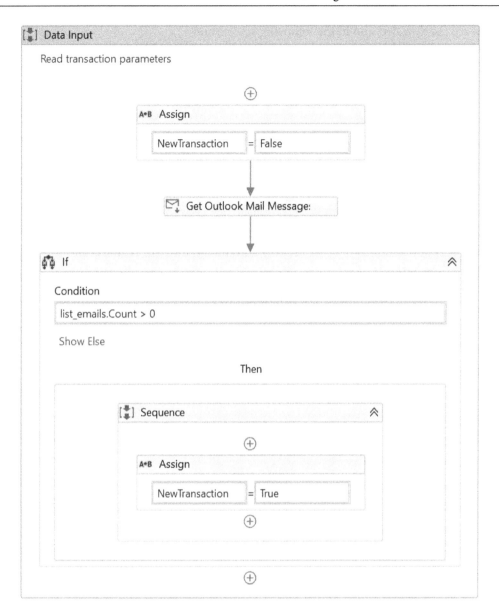

Figure 9.20 – Data Input sequence

Looking back at the **Data Input** sequence, the automation will fetch Outlook mail messages, outputting its result into the `list_emails` variable. If `list_emails.count` is greater than 0, then there are emails within the `Email Classification` folder, and automation can set **NewTransaction** equal to `True`.

You can see the flow decision in the following screenshot:

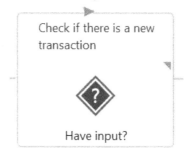

Figure 9.21 – Have input? flow decision

By setting **NewTransaction** equal to `True`, the **Have input?** flow decision within the **Main** workflow will return `True`, leading automation into the **Transaction Processing** flowchart.

In the **Main** workflow, let's replace the **Transaction Processing** flowchart with a sequence named **Classify Email**. Your screen should now look like this:

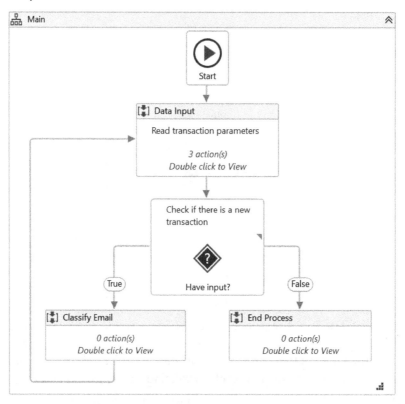

Figure 9.22 – Main workflow

The reason for changing the **Transaction Processing** section into a sequence instead of a flowchart is that the classification of an email is very sequential, with few business variations.

Let's build out the sequence, as follows:

1. Double-click on **Classify Email** to open the sequence.

2. Drag an **Invoke Workflow File** activity into the workflow, with the following parameter:

    -   **Workflow file name**: `Classify_Email.xaml`

    The purpose of the **Invoke Workflow File** activity is to call our ML workflow created earlier. Within the activity, we need to import arguments from the main file to the `Classify_Email` workflow.

3. Click **Import Arguments**, as illustrated in the following screenshot:

Figure 9.23 – Invoke Workflow File before importing arguments

Once **Import Arguments** is clicked, a popup will appear showcasing the arguments we added to `Classify_Email`.

4. Set the following arguments:

    -   `in_EmailText`: `list_emails(0).Body`

    -   `out_Classification`: `email_classification`

    -   `out_Confidence`: `email_confidence`

The preceding arguments are shown in the following screenshot:

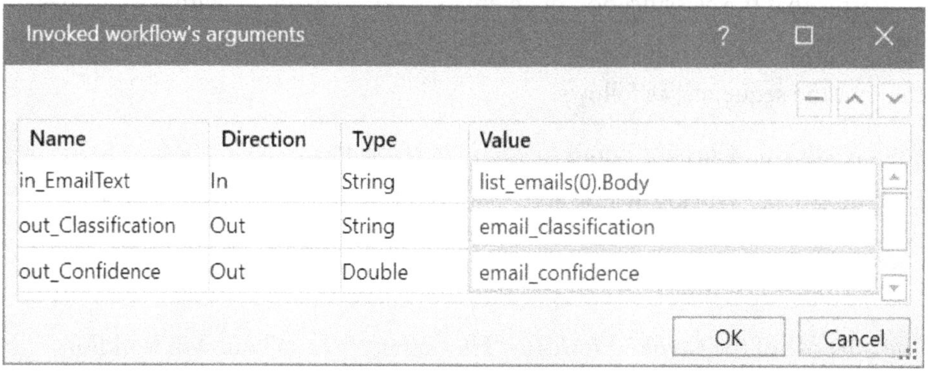

Name	Direction	Type	Value
in_EmailText	In	String	list_emails(0).Body
out_Classification	Out	String	email_classification
out_Confidence	Out	Double	email_confidence

Figure 9.24 – Arguments added

Because list_emails is a list of mail messages from the **Get Outlook Mail Messages** activity, we need to index the 0th email with (0), then pass the email's body with .body. email_classification and email_confidence are two new variables that we need to create to hold the output of Classify_Email.

Once the **Invoke Workflow File** activity is set up, the next step of the **Classify Email** sequence is to check the confidence output by the ML skill. In this case, any confidence that is less than 75% we'll move to the Low Confidence folder.

5. Drag an **If** activity into the workflow, with the following condition:

- email_confidence > .75

6. Within the **Else** block of the **If** activity, drag an **Invoke Workflow File** activity into the workflow, with the following parameters:

- **Workflow file name**: Move_Email.xaml

As with the Classify_Email **Invoke Workflow File** activity, for this Move_Email file activity, we want to move the low-confidence email into our Email Classification\Low Confidence folder.

7. Click **Import Arguments** and set the following arguments:

- in_MailMessage: list_emails(0)

- in_Folder: "Inbox\Email Classification\Low Confidence"

With the `Move_Email` workflow added, our **Classify Email** sequence should look like this:

Figure 9.25 – Classify Email sequence

Currently, our workflow can classify our email, check its confidence, and move the email to our `Low Confidence` folder if `email_confidence` is less than 75%. However, we still need to add logic when `email_confidence` is greater than or equal to 75%.

8. Within the **Then** block of the **If** activity, drag a **Switch** activity into the workflow, setting the following parameters:

    - **Expression**: email_classification
    - **Type**: String

    Inside the Switch statement, we're going to add two cases, one for Spam and one for Ham. Spam classifications are going to be moved to the Spam folder, while Ham classifications are going to be moved to the Invoices folder.

9. Click **Add a New Case** and name that case spam.

10. Within the case, drag an **Invoke Workflow File** activity into the workflow, with the following parameter:

    - **Workflow file name**: Move_Email.xaml

11. Click **Import Arguments** and set the following arguments:

    - in_MailMessage: list_emails(0)
    - in_Folder: "Inbox\Email Classification\Spam"

    Hopefully, you should see the benefit of creating a supplemental Move_Email workflow now. In the Ham case, let's perform the same steps to move a Ham email into the Invoices folder.

12. Click **Add a New Case** and name that case ham.

13. Within the case, drag an **Invoke Workflow File** activity into the workflow, with the following parameter:

    - **Workflow file name**: Move_Email.xaml

14. Click **Import Arguments** and set the following arguments:

    - in_MailMessage: list_emails(0)
    - in_Folder: "Inbox\Email Classification\Invoices"

By adding the `Ham` case to our `Switch` statement, our **Classify Email** sequence should look similar to this:

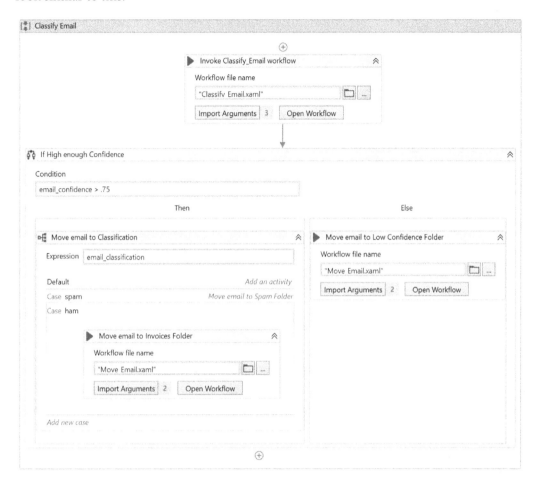

Figure 9.26 – Classify Email workflow

That concludes configuring the **Classify Email** sequence and automation workflows of our use-case project. In this section, we created our UiPath project and built our ML and supplemental workflows. We then connected our supplemental workflows together within the `Main.xaml` file of our project. In the next section, we will test our development and add additional functionalities to close the feedback loop and retrain our ML skill.

Testing to ensure stability and improve accuracy

With the initial development of the use case complete, we can venture into testing out how well our automation performs. Testing any automated workflow before deployment is crucial in order to ensure the automation works as expected. In this section, we will test with sample data, as well as adding retraining functionality to our automation to retrain the ML skill.

Testing with sample emails

To test out the functionality of our use case, we must prepare our inbox with sample email messages for automation to run through. In *Chapter 7, Testing and Refining Development Efforts*, we discussed that when training cognitive automation, we should have multiple datasets to test our ML skill, as outlined here:

- **Training dataset**: Initial dataset to train the ML model

- **Evaluation dataset**: A smaller set of data used to evaluate the training of the ML model

- **Testing dataset**: An unbiased dataset used to expose the trained ML model to real-life data

Since we have already performed the initial training and evaluation, we only need to test our automation with our unbiased data.

To start testing, navigate to the following link to retrieve test emails:

```
https://github.com/PacktPublishing/Democratizing-Artificial-
Intelligence-with-UiPath/tree/main/Chapter09/Email%20Data/
Test%20Emails
```

At the preceding link, you will find emails in the .msg format that you can move into Outlook to test with. If you don't have Outlook, download the example project from the following link:

```
https://github.com/PacktPublishing/Democratizing-Artificial-
Intelligence-with-UiPath/tree/main/Chapter09/Email%20
Classification
```

Within the example UiPath project is a workflow, **Create Test Emails**, that can be used to generate emails. **Create Test Emails** will read a **comma-separated values (CSV)** file, and for each row in the CSV file to create an email. Prior to running, set the two arguments of the workflow, as follows:

- `in_DataFile`: Set to the file path of where your data CSV resides

- `in_To`: Set to your email address to send an email to

The two arguments are shown in the following screenshot:

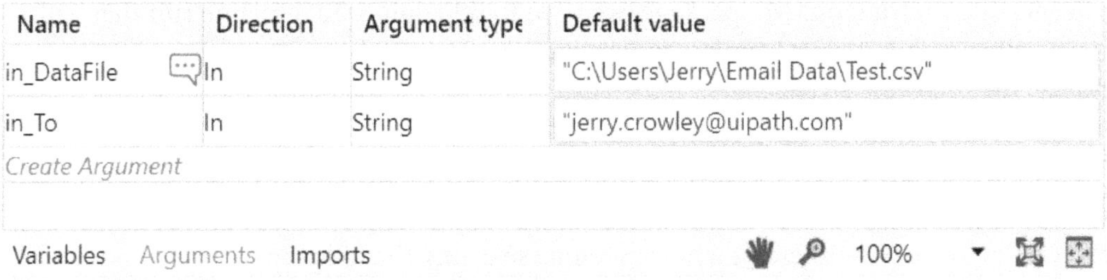

Name	Direction	Argument type	Default value
in_DataFile	In	String	"C:\Users\Jerry\Email Data\Test.csv"
in_To	In	String	"jerry.crowley@uipath.com"
Create Argument			

| Variables | Arguments | Imports | | | 100% | | |

Figure 9.27 – Setting Create Test Emails arguments

> **Important Note**
>
> If you don't have Outlook, you may need to change the **Send Outlook Mail Message** activity to an activity that works with your email provider.

Feel free to test with a subset of 48 messages, or to individually test with all messages. If you notice that your results aren't as expected, then you may want to rerun the training and evaluation pipelines with the full dataset. In the next section, we'll look into improving performance by creating a retraining workflow to retrain our ML skill for us.

Building a retraining workflow

To improve the performance of our ML skill over time, we can close the feedback loop—essentially building the capability to provide feedback to our ML skill and train the skill as new data comes in.

Adding retraining capabilities to our workflow requires the following two steps:

1. Building a full pipeline that can be leveraged to retrain.

2. Building an automation workflow to upload new data to our full pipeline.

Once these two steps are completed, our automation can effectively retrain itself based on the feedback a human provides.

Building a full pipeline

When we built the ML skill earlier in the chapter, we ran a training and evaluation pipeline to get our ML skill ready for automation. While we previously only ran these pipelines once, in a closed feedback loop model we need to continuously rerun these pipelines to train and evaluate our ML package over time.

So, to effectively retrain our ML package, we need to schedule a full pipeline run that will run a pipeline with both our training and evaluation datasets. To create a full pipeline, complete the following steps:

1. Navigate back to our `Email Classification` project in AI Center.

2. On the **Pipelines** page, click **Create New** to create a new pipeline.

3. Click **Create New** to create a new pipeline, selecting the following attributes:

    -   **Pipeline type**: `Full run`

    -   **Choose package**: `TextClassification`

    -   **Choose package major version**: `6`

    -   **Choose package minor version**: `0`

    -   **Choose input dataset**: `Email Classification/Train`

    -   **Choose evaluation dataset:** `Email Classification/Evaluation`

    -   **Enable GPU**: Turn this off

    -   **Schedule**: **Recurring, weekly**

4. Under **Environment Variable**, add the following environment variables:

    -   `input_column`: `text`

    -   `target_column`: `label`

Your screen should now look like this:

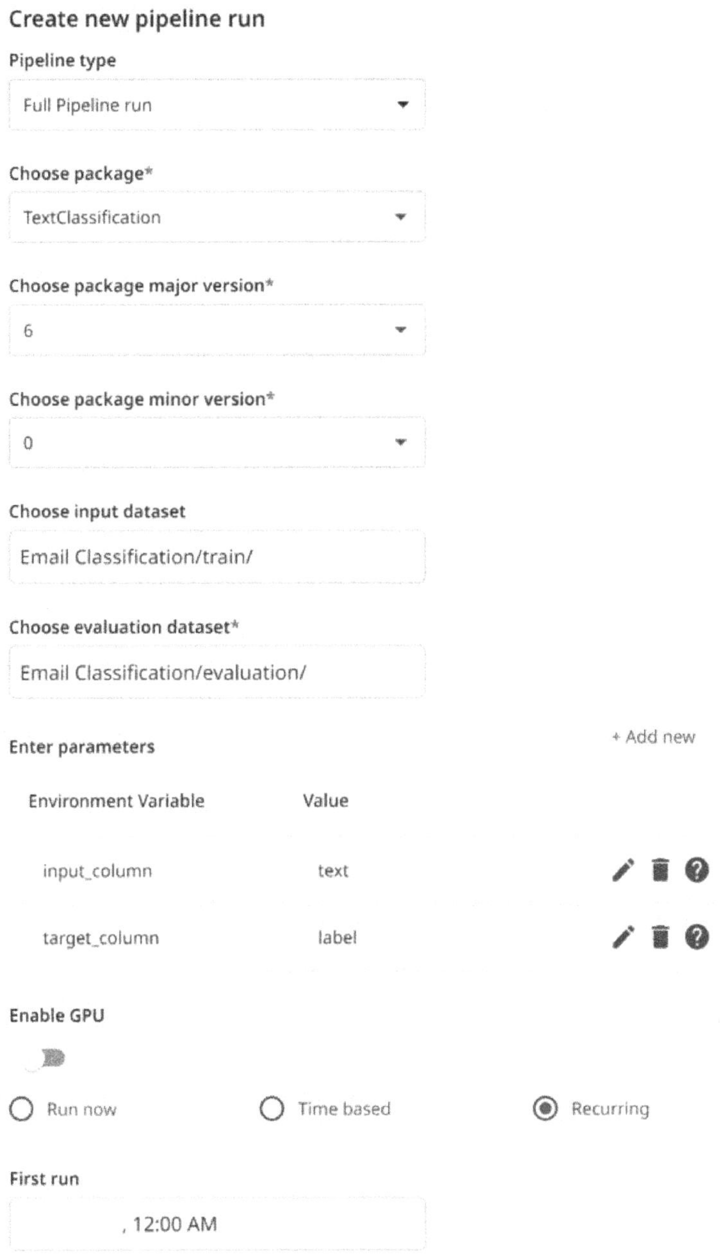

Create new pipeline run

Pipeline type

Full Pipeline run

Choose package*

TextClassification

Choose package major version*

6

Choose package minor version*

0

Choose input dataset

Email Classification/train/

Choose evaluation dataset*

Email Classification/evaluation/

Enter parameters   + Add new

Environment Variable	Value
input_column	text
target_column	label

Enable GPU

○ Run now   ○ Time based   ◉ Recurring

First run

, 12:00 AM

Figure 9.28 – Creating a training pipeline

5. Click **Create** to start the training pipeline.

> **Important Note**
> Make sure to choose package minor version 0 for retraining.

With the **First run** parameter set to run every 7 days, the full pipeline is set to run with new datasets. In the next section, we'll look into updating our datasets with validated data.

Building an automation workflow

In order to close the feedback loop, we need a way for a human to validate the automation's performance. With Document Understanding, we could use Validation Station to retrain a classifier and extractor; however, in this use case, we'll use email folders for the end user to move incorrect emails into.

With the `Retrain-Not Spam` and `Retrain-Spam` folders, an end user can move an incorrectly classified spam email into the `Not Spam` folder, and vice versa. Automation can then look into these folders and add the email to the training set, retraining our ML skill over time.

The first step is to build our retraining workflow. The objective for the retraining workflow is to add each email from the `Retrain-Not Spam` and `Retrain-Spam` folders to a singular CSV file, move each email to its destination email folder, and then upload the CSV file to AI Center to be added to our training dataset. Follow these next steps:

1. In the `Email Classification` project, create a new workflow named `Retrain_Model`.

2. Create two input parameters, as follows:

    *   `in_listSpamEmails: List<MailMessage>`

    *   `in_listNotSpamEmails: List<MailMessage>`

    The parameters are shown in the following screenshot:

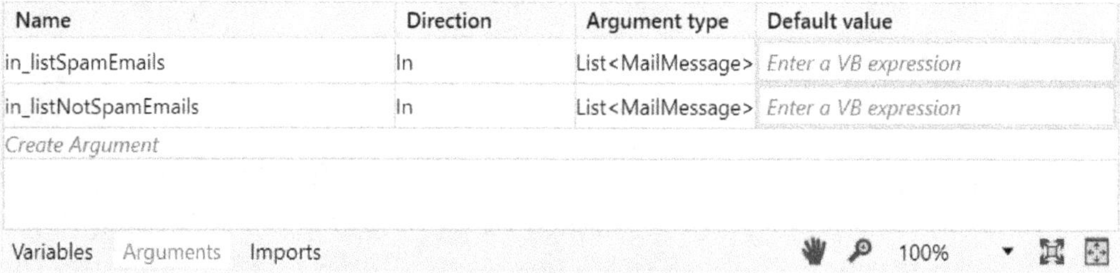

Name	Direction	Argument type	Default value
in_listSpamEmails	In	List<MailMessage>	Enter a VB expression
in_listNotSpamEmails	In	List<MailMessage>	Enter a VB expression
Create Argument			

| Variables   Arguments   Imports | | | 🖐 🔍 100% ▼ 🔁 🔲 |

Figure 9.29 – Retail_Model arguments

3. Drag a **Build Datatable** activity into the workflow, outputting to a new `retrain_`
   `DT` variable and building two columns, as follows:

   - `text: String`
   - `label: String`

   You can see what the data table looks like in the following screenshot:

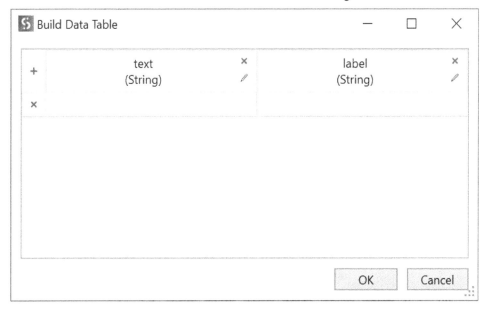

Figure 9.30 – retrain_DT data table

4. Drag a **For Each** activity into the workflow, looping through each email in `in_`
   `listSpamEmails`.

5. Within the **For Each** activity, drag an **Add Data Row** activity into the workflow,
   setting the following parameters:

   - `ArrayRow: {email.Body, "spam"}`
   - `DataTable: retrain_DT`

6. Drag a **Move Outlook Mail Message** activity into the workflow, setting the
    following parameters:

    - `MailMessage: email`

    - `MailFolder: "Inbox\Email Classification\Spam"`

    - The parameters are shown in the following screenshot:

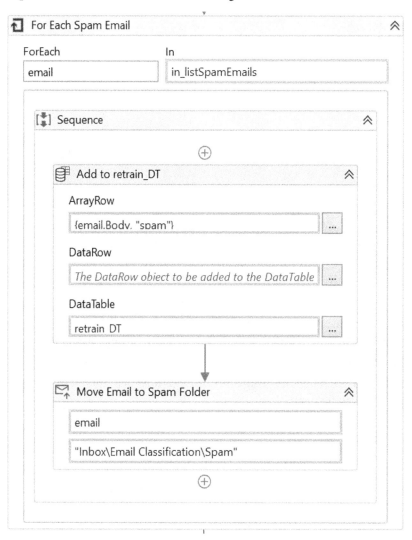

Figure 9.31 – For Each Spam Email

With the `for` loop, adding each spam email to our data table, and moving each spam email into the `Spam` folder, we can add the same steps for the not-spam emails:

1. Drag a **For Each** activity into the workflow, looping through each email in `in_listNotSpamEmails`.

2. Within the **For Each** activity, drag an **Add Data Row** activity into the workflow, setting the following parameters:

    -   ArrayRow: {email.Body, "ham"}
    -   DataTable: retrain_DT

3. Drag a **Move Outlook Mail Message** activity into the workflow, setting the following parameters:

    -   MailMessage: email
    -   MailFolder: "Inbox\Email Classification\Invoices"

With both email folders traversed through, added to `retrain_DT`, and moved into their intended folders, we can now focus on writing `retrain_DT` to a CSV file and uploading the CSV file to our AI Center training dataset:

1. Drag a **Write CSV** activity into the workflow, setting the following parameters:

    -   FilePath: "retrain.csv"
    -   DataTable: retrain_DT
    -   **Add Headers**

2. Drag an **Upload File** activity into the workflow, setting the following parameters:

    -   **Project**: Email Classification
    -   **Dataset**: Email Classification
    -   **Path into dataset**: "train/"
    -   **File**: "retrain.csv"

The process is illustrated in the following screenshot:

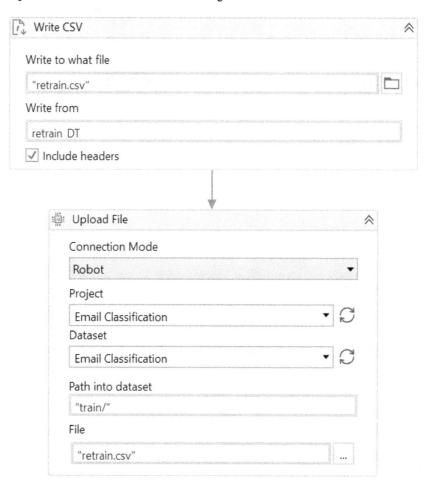

Figure 9.32 – Uploading retrain_DT to our training dataset

With `Retrain_Model.xaml` complete, we can focus on connecting it with `Main.xaml` and fetching emails from the `Retrain-Spam` and `Retrain-NotSpam` folders:

1. Open `Main.xaml` and open the **End Process** sequence.

2. Drag a **Get Outlook Mail Messages** activity into the workflow, setting the following parameters:

    -   `MailFolder:` `"Inbox\Email Classification\Retrain-Spam"`

    -   **Top**: `-1`

    -   **Output**: `list_retrainSpamEmails`

3. Drag another **Get Outlook Mail Messages** activity into the workflow, setting the following parameters:

    -   `MailFolder:` `"Inbox\Email Classification\Retrain-NotSpam"`

    -   **Top**: `-1`

    -   **Output**: `list_retrainNotSpamEmails`

    With our emails extracted, we can now pass them to our newly created `Retrain_Model` workflow, but we only need to run this workflow if we have any emails.

4. Drag an **If** activity into the workflow, setting the condition to `(list_retrainSpamEmails.Count > 0) OR (list_retrainNotSpamEmails.Count > 0)`.

5. Within the **If** activity, drag an **Invoke Workflow File** activity into the workflow, setting the following parameters:

    -   **Workflow file name**: `"Retrain_Model.xaml"`

    -   `in_listSpamEmails:` `list_retrainSpamEmails`

    -   `in_listNotSpamEmails:` `list_retrainNotSpamEmails`

The process is illustrated in the following screenshot:

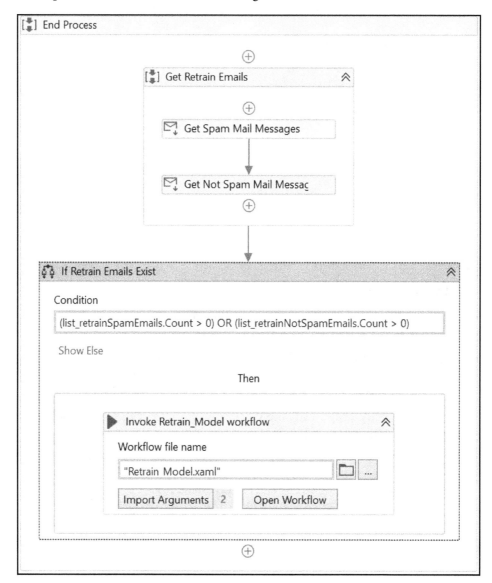

Figure 9.33 – End Process sequence of Main.xaml

And that's it! Within *Figure 9.33*, we can see what our **End Process** sequence should resemble. In this section, we successfully added retraining capabilities to our automation project by first building a full pipeline to retrain our ML package every 7 days and then building an automation workflow to extract validated emails and add the validation to our training pipeline. With the development and testing portions complete, in the next section, we will venture into deploying the automation into UiPath Assistant.

Deploying with the end-user experience in mind

With the initial development and testing of the use case complete, let's venture into deploying the automation into a production environment. In this section, we will work on deploying the automation into UiPath Assistant so that it can be run in an attended scenario.

Deploying the project

In a production setting, we shouldn't run our automation directly from UiPath Studio, nor should we expect our end users to have access to UiPath Studio. Thus, we must deploy our automation into UiPath Orchestrator to allow others to run our automation from UiPath Assistant.

> **Important Note**
>
> More information on deploying automation can be found at the following link:
>
> ```
> https://docs.uipath.com/studio/docs/about-
> publishing-automation-projects
> ```

For this use case, we're going to quickly publish the automation into our own personal workspace within UiPath Assistant. Follow the next steps:

1. Within the Email Classification project, click the **Publish** button in the **Design** ribbon of UiPath Studio.

2. Check that the **Publish to** field is set to **Orchestrator Personal Workspace Feed**, as illustrated in the following screenshot:

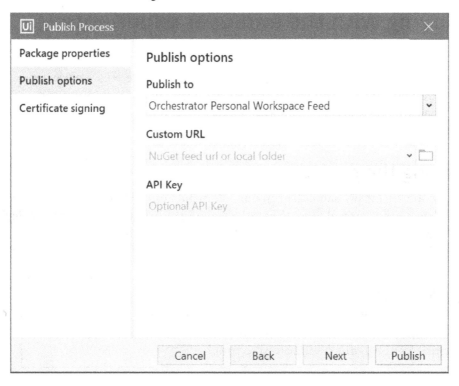

Figure 9.34 – Publishing the automation

3. Click **Publish**.

Once published, you should be able to see the automation within UiPath Assistant. It should look like this:

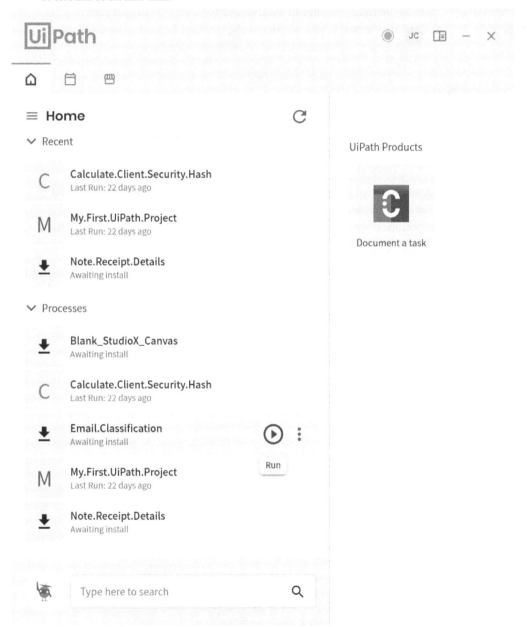

Figure 9.35 – Project deployed to UiPath Assistant

Clicking the **Play** button to the right of `Email.Classification` will install the automation on your machine and start execution.

Once deployed, you are free to run the automation directly from UiPath Assistant. Any changes to the code can be made in UiPath Studio and then republished following the steps just discussed. In this section, we completed the use case build by deploying the `Email Classification` project into UiPath Assistant—great job!

Summary

In this chapter, we covered building a cognitive automation use case with *UiPath AI Center*. In the construction of the use case, we learned how to create a project in AI Center, load training and evaluation datasets, build training and evaluation pipelines, and—finally—deploy an ML skill. We also learned how to leverage the ML skill within our UiPath project and learned how to close the feedback loop with UiPath's **Upload File** activity, allowing us to upload training data back into the training dataset of AI Center. With these foundations, you can now venture off and build your own cognitive automation.

In the next chapter, we will take another hands-on approach by working together to build the last of three use cases with UiPath and Druid. We will learn how to integrate chatbots and UiPath.

10
Use Case 3 – Chatbots with Druid

We can bridge conversational **artificial intelligence** (**AI**) and UiPath automation with intelligent chatbots to modernize operations and enhance the **user experience** (**UX**). With chatbots, we can integrate intelligent communication tools into messaging apps or chat windows, giving end users an almost immediate resolution to their inquiries. With chatbots and UiPath, we can go one step further in fulfilling the requests of customers—not only can we respond to customer inquiries using intelligent chatbots, but we can also launch automation processes to provide **end-to-end** (**E2E**) fulfillment of customer requests.

In this chapter, we will work together to build a chatbot that can interact with human users and launch UiPath automations. As we build the use case, we will follow a similar development life cycle as outlined in the previous chapters. We will start with understanding the current state of our automation opportunity, and then design the future state of the use case with automation involved. Once the future state is designed, we will build the chatbot with Druid and the automation with UiPath Studio.

In this chapter, *Use Case 3 – Chatbots with Druid*, we will cover the following topics:

- Understanding the current state of the use case
- Creating a future state design
- Building a solution with Druid
- Building an automation solution
- Testing to ensure stability and improve accuracy
- Considerations for production

Once complete, we will have built a chatbot with Druid with a connection to UiPath automation.

Technical requirements

All code examples for this chapter can be found on GitHub at the following link: `https://github.com/PacktPublishing/Democratizing-Artificial-Intelligence-with-UiPath/tree/main/Chapter10`.

Working with UiPath is very easy, and with the Community version, we can get started for free.

For this section, you will need the following:

- UiPath Enterprise Cloud
- UiPath Studio 2021.4+
- Druid Oxygen v1.57.0+

> **Important Note**
> Trial licenses for Druid can be acquired at `https://www.druidai.com/license-activation`.

Check out the following video to see the Code In Action at: `https://bit.ly/3rt4C5x`

Understanding the current state of the use case

Before we venture into designing and building a use case, we need to understand the current state of the opportunity. With an increase in hiring in *Company ABC* has come an influx of service requests to the **information technology** (**IT**) team. Some of these service requests are simple requests that can be resolved with little human interaction, such as the following:

- Resetting passwords

- Upgrading colleagues' video conferencing software to the professional version

With the increased hiring of *Company ABC*, a lot of the support representative's time is spent resolving quick, simple requests. To increase the capacity of the support team, the IT team has been asked to automate these two requests.

Given that the tasks of resetting passwords and upgrading application licenses are straightforward, we can see that this automation opportunity has the following characteristics:

- **Repetitive in nature**: Support representatives must perform the same series of steps for each service request.

- **Time-consuming**: Completing each service request can take time away from support representatives performing more meaningful tasks.

- **Rules-based and structured**: Resetting a user's password or upgrading their license type is very rule-based.

Automation of these tasks can provide a great deal of impact for not only the IT team but also for colleagues within *Company ABC*. In the next section, we will focus on creating a future state design for implementing this IT helpdesk bot.

Creating a future state design

With our new understanding of the current state and the pain points of the IT team, we can start by creating a future state design of the automation, as represented in the following diagram:

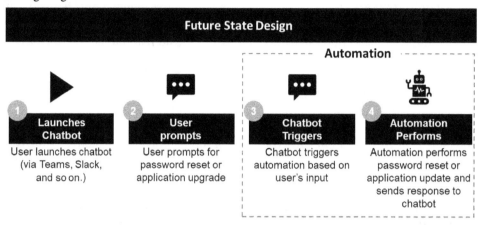

Figure 10.1 – Future state design of the IT helpdesk bot

In the future state design of the use case, we could have a human launch our Druid chatbot (which could be connected to Slack, Microsoft Teams, or via the web), and prompt the chatbot to help with either password resets or application upgrades. Once the chatbot receives the request, it can trigger UiPath automation with an input argument on the specific task (password reset or application upgrade).

Before we go into details on the future state design, it's important to know some of the basics of building a chatbot. All conversational flows contain intent and dialog, which are the building blocks of a chatbot. When we represent a conversation between a person and our chatbot, the dialog between the two will be based on some intent. This is outlined further here:

- **Intent**: A purpose for starting the conversation (also known as an utterance)

- **Dialog**: A conversation that is in response to an intent

- **Flows**: A combination of intents and dialog

All Druid chatbots are based on this methodology, whereby a user initiates a conversation with a chatbot (an utterance). Based on that utterance, the chatbot best matches with an intent that then starts a dialog, which could contain free text, buttons, or other inputs.

With this understanding, we can design the flows of our chatbot. By default, a chatbot should have the following flows configured:

- **Welcome flow**: A flow to welcome the user once they initiate the chatbot
- **Fallback flow**: A flow for the chatbot to fall back on if the bot does not know the answer or response to a user's input

In addition to leveraging the Welcome and Fallback flows, we will need to add one additional flow for our automation. This flow will be called `UiPath_Flow`, where we will provide a set of buttons for the user to choose from. Depending on the button the user chooses, Druid will trigger an automation with an argument to either reset their password or to upgrade their application license.

That concludes our future state design. In this section, we discussed chatbot basics, created a diagram to resemble our future state design, and introduced the three flows we will build in our chatbot. In the next section, we will get started with development by building the Druid chatbot.

Building a solution with Druid

As we venture into building our defined use case, we'll start with building our chatbot component with Druid. For this use case, we'll start out with Druid's **Druid Starter** template to set the foundation of our chatbot. This template comes with a definition of system entities and common flows that every chatbot should leverage.

Creating a project

To start a project with the **Druid Starter** template, complete the following steps:

1. In the main menu, click on the **Bot** dropdown and select **AddNewBot**.

2. Once the **Create New Bot** window appears, enter IT Help Desk Bot as the bot's name, choose English (United States) as the default language, and click **Save**. The process is illustrated in the following screenshot:

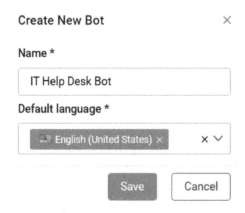

Figure 10.2 – Creating a new bot

Now that the bot is created, let's import the **Druid Starter** template to get us started.

3. In the main menu, click on the **Solution** dropdown and select **Import Solution**.

4. Once the **Solution Library** page appears, select **Druid Starter** and then select **Import Solution**.

With the **Druid Starter** template installed, we have a good foundation for us to build off of. Let's get started with creating some conversational flows.

Editing the Welcome flow

The first flow of the **Druid Starter** template is known as the **Welcome** flow. The purpose of this flow is to welcome the user when they first interact with a chatbot. If you navigate to the flows of the project (click **Flows** under **Solution Contents**), you will notice our Welcome flow, named first-welcome-flow, as illustrated in the following screenshot:

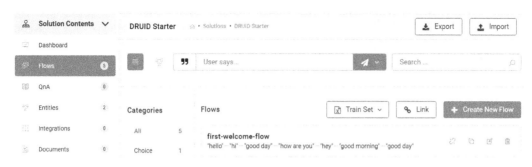

Figure 10.3 – Welcome flow

Within `first-welcome-flow`, let's add our first prompt asking the user for their email address, as follows:

1. Within `first-welcome-flow`, click the **Edit** button to edit the flow.

2. Click **AddChildStep** within the parent message, as illustrated in the following screenshot:

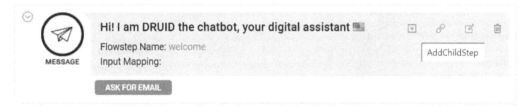

Figure 10.4 – Adding a child step

3. Once **newMessage** appears, click the **Edit** button to edit the child step, setting the following attributes:

    *   **Step name**: `Ask for Email`

    *   **Type**: `Prompt`

    *   **Step message**: `Please input your email address`

    *   **Input mapping**: `@email`

The process is illustrated in the following screenshot:

EditFlowStep   ✕

                                                              Save   Cancel

ⓘ   **General**   ⌄

Step name

| Ask for Email |

Type

| Prompt   ⌄ |

🖼 Step message

| B *I* H ⌂ ☺ ⚲ 🖼 ☰ ☷ </> “ Q Preview   ⤢ |
| Please input your email address |

Input mapping   SkipIfKnown

| @email |

Figure 10.5 – Editing newMessage

By creating a prompt as a child step beneath the welcome message, our chatbot
will now always ask for a user's email after it welcomes the user. By creating
an @email input mapping, our chatbot will also save the user's email address
as an @email variable.

After we prompt for an email address, we should add one more message confirming
the user's email before transitioning to another flow asking for their issue.

4. Within our newly created Ask for Email step, click **AddChildStep**.

5. Once **newMessage** appears, click the **Edit** button to edit the child step, setting the
    following attributes:

    - **Step name**: Confirm Email

    - **Type**: Flow

    - **Step message**: Thanks, @email!

    With the Confirm Email step added, first-welcome-flow should look
    similar to this:

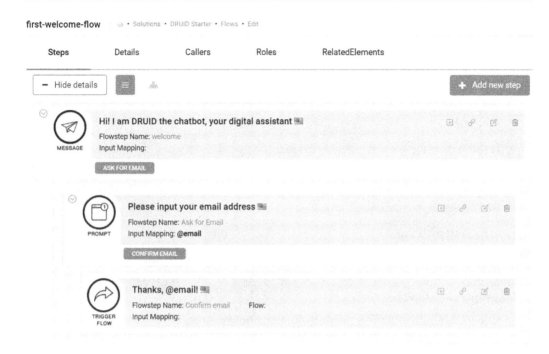

Figure 10.6 – first-welcome-flow

For now, that concludes the welcome flow. As we build out the other flows, we'll come back and make some small edits.

Creating an automation flow

With our welcome flow complete, let's venture into the main flow of the chatbot: the automation flow. In this flow, we'll present the user with options to choose from, launching a UiPath automation based on the user's input. Let's get started, as follows:

1. Click **Create New Flow** to create a new flow, setting the following attributes:

    - **Name**: UiPath_Flow

    - **Description**: Main Flow for IT Help Desk Bot

    - **Utterances**: help

    - **Category**: Automation Flow

    With the flow created, let's add a choice step to the flow, giving users the ability to choose between resetting their password or upgrading their license type.

2. Click **+ Add new step** to create a new step, setting the following attributes under the **General** tab:

    - **Step name**: `Self-Help`
    - **Type**: `Choice`
    - **Step message**: `What can I help you with?`
    - **Input mapping**: `@self_help_choice`

3. Within the **Metadata** tab of the `Self-Help` step, set the following attributes:

    - **BlockUserInput**: `True`
    - **Is first step?**: `True`

    Additionally, within the **Metadata** tab, you'll see an option to add choices (see *Figure 10.7*). These choices will be button options for the user to choose from when they interact with the chatbot.

4. Click **+AddChoice**, setting the following attributes:

    - **Caption**: `Password Reset`
    - **Value**: `pass_reset`

5. Click **+ AddChoice**, setting the following attributes:

    - **Caption**: `Upgrade AppY`
    - **Value**: `upgrade_appY`

    The following screenshot depicts the process of adding choices:

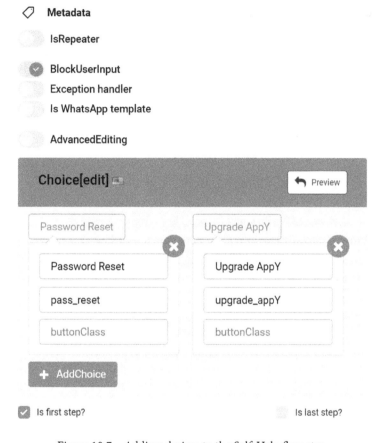

Figure 10.7 – Adding choices to the Self-Help flow step

6. Click **Save** to save your choices.

Great! By clicking **Save**, we have added a choice that will look similar to this when we test it out. Note that by checking **BlockUserInput**, the user can only select from the two buttons and not add free text:

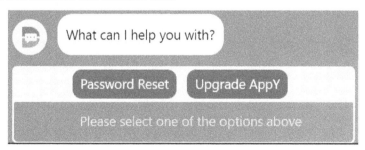

Figure 10.8 – Druid choices

With our choice step added, the next step is to add a UiPath connection.

7. Use Druid's **Flow Diagram Designer** by clicking on the **Flow** button highlighted in the following screenshot:

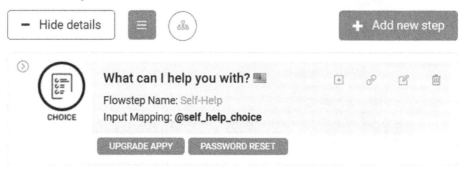

Figure 10.9 – Enabling Druid's Flow Diagram Designer

Once the **Flow Diagram Designer** appears, you should be able to see the Self-Help step we just added. Hovering over the + sign underneath the flow step will present several options to add as a child step, as illustrated in the following screenshot:

Figure 10.10 – Adding a child step

8. Add **UiPath Automation** by clicking on the **Ui** button.

9. Within the `Launch Automation` step, set the following attributes:

   - **Step name**: `Launch Automation`

   - **Type**: `UiPath`

   - **Technology**: `Automation: attended (Robot.JS)`

   - **Manually set process information**

   - **Process**: `IT_Helpdesk_Chatbot`

   - **Step message**: `Launching UiPath Automation...`

10. Within `Launch Automation`, set the following process **input/output (I/O)** arguments:

    - **Name**: `in_Task`; **Direction**: `In`; **Druid entity name**: `@self_help_choice`

    - **Name**: `in_Email`; **Direction**: `In`; **Druid entity name**: `@email`

    - **Name**: `out_SRNumber`; **Direction**: `Out`; **Druid entity name**: `@SR_Number`

    The process is illustrated in the following screenshot:

Figure 10.11 – Launch Automation step details

In subsequent sections, we will create an `IT_Helpdesk_Chatbot` UiPath automation with the arguments listed in *Figure 10.11*.

Once the UiPath automation is added, you will notice that Druid will add multiple child steps, as follows:

- **Request Successful**: Message displayed by the chatbot when it successfully sends UiPath a request to run an automation

    - **Error sending the request**: Error message displayed by the chatbot when it unsuccessfully sends UiPath a request to run an automation

- **After Job Complete (Success)**: Message displayed by the chatbot when the UiPath automation job completes successfully

- **After Job Complete (Error)**: Error message displayed by the chatbot when the UiPath automation job completes successfully

In the **After Job Complete (Success)** message, let's change the message text to include a **service request number** (**SRN**) that will be output by the UiPath automation.

11. Open the **After Job Complete (Success)** step and set the following attributes:

- **Step message**: `Success, job complete. Your request number is @SR_Number for reference`

Once the UiPath automation ends, we want the chatbot to prompt the user, asking whether more assistance is needed. Let's add another choice after the **After Job Complete (Success)** flow step.

12. Add a choice after the **After Job Complete (Success)** flow step, setting the following attributes:

- **Step name**: `Ask for more Help`
- **Type**: `Choice`
- **Step message**: `Is there anything else I can help you with?`
- **Input mapping**: `@more_help`
- **BlockUserInput**: `True`
- **Choice1**: `Yes`
- **Choice2**: `No`

The process is illustrated in the following screenshot:

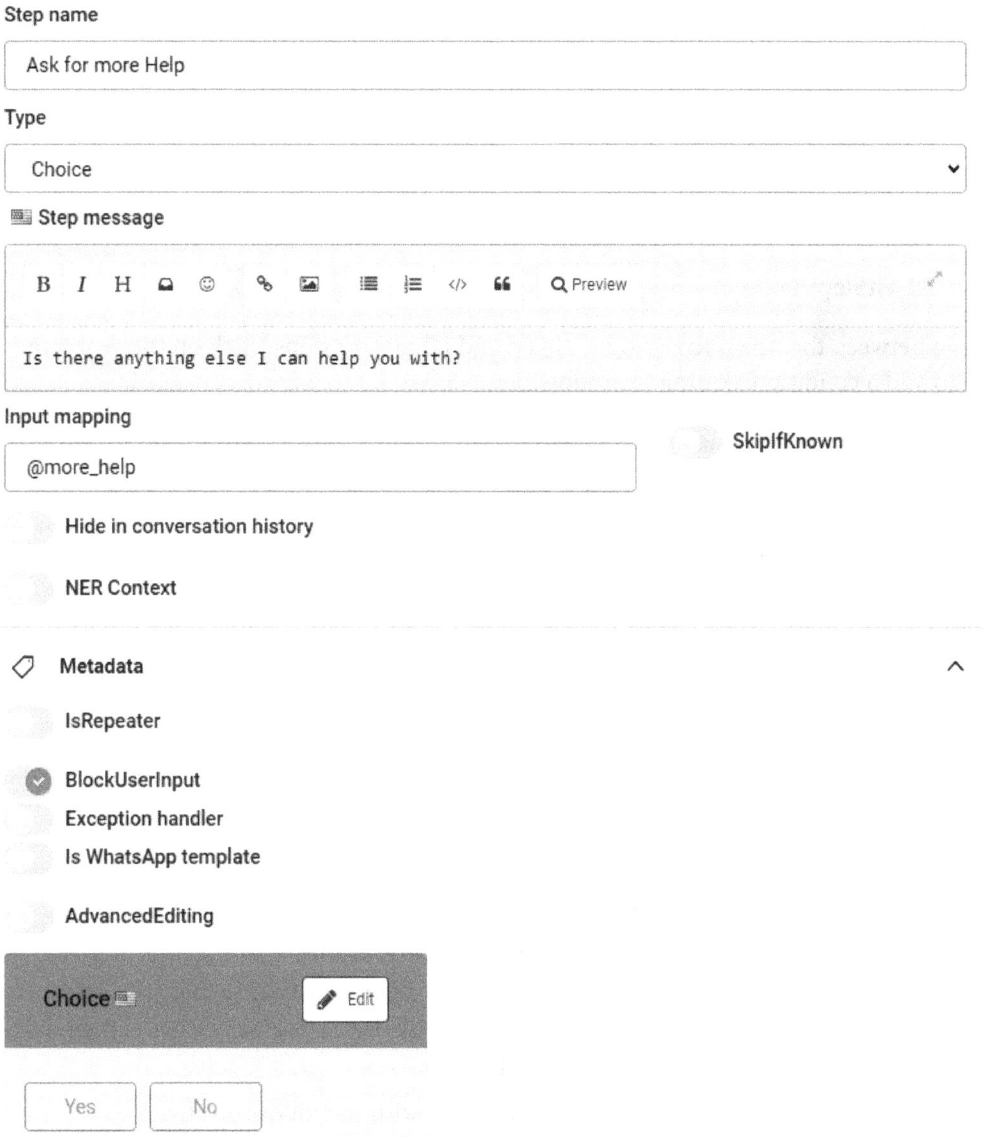

Step name

Ask for more Help

Type

Choice

Step message

| B | *I* | H | ☐ | ☺ | ⚙ | ▣ | ☰ | ☷ | </> | 66 | Q Preview |

Is there anything else I can help you with?

Input mapping

@more_help

SkipIfKnown

Hide in conversation history

NER Context

◇ **Metadata** ⌄

IsRepeater

BlockUserInput

Exception handler

Is WhatsApp template

AdvancedEditing

Choice ▣ | ✏ Edit

| Yes | No |

Figure 10.12 – Ask for more Help choice

13. Within **Flow Diagram Designer**, make sure to connect both the **Error sending the request** and **After Job Complete (Error)** flow steps to the newly created `Ask for more Help` choice.

After `Ask for more Help`, we should have one final message, either thanking the user if they say no to any more help, or navigating them back to the beginning of the flow if they do ask for more help.

14. After `Ask for more Help`, add a child step with the following attributes:

- **Step name**: `Closing`

- **Type**: `Message`

- **Message**: `Okay! Have a great day`

- **IsLastStep**: `True`

15. In between the `Ask for more Help` and `Closing` steps, click on the gear icon to add a condition, adding the following:

- **Condition**: `@more_help=="No"`

The following screenshot illustrates this:

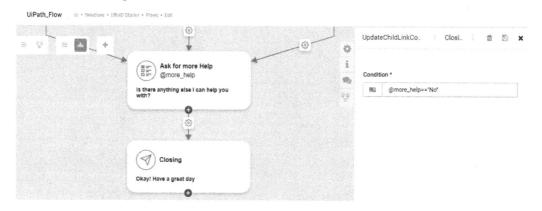

Figure 10.13 – Adding a condition to Closing

The purpose of this condition is to only route the conversation to the `Closing` step if the user's choice was `No` for additional help. If the user selects `Yes`, then they should be rerouted to the beginning of the flow.

16. After `Ask for more Help`, add a child step with the following attributes:

- **Step name**: `More Help`

- **Type**: `Message`

- **Message**: `Okay!`

17. In between the `Ask for more Help` and `More Help` steps, click on the gear icon to add a condition, adding the following:

- **Condition**: `@more_help=="Yes"`

18. Drag a connecting arrow from the `More Help` step to the `Self-Help` step, as illustrated in the following screenshot:

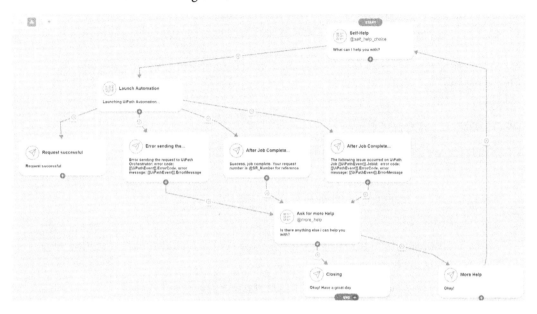

Figure 10.14 – Completed UiPath_Flow process

With the automation flow complete, we have one last step before we can move on to building the UiPath automation. We need to connect the Welcome flow with the UiPath flow.

19. Navigate back to `first-welcome-flow`.

20. Edit the `Confirm email` step, setting the following attributes:

- **Subflow**: `UiPath_Flow`

And that's it! We have successfully created a chatbot flow with Druid that connects to UiPath. In subsequent sections, we will build the UiPath automation component of this use case, and then test out the functionality.

Building an automation solution

With our chatbot built, we can shift our focus to creating automation workflows for the use case. Approaching this use case, we will build our project in a couple of steps, as follows:

1. Creating a project
2. Building `Reset_Password` and `Upgrade_App` workflows
3. Building the **Main** workflow

After completing these steps, we will have the foundation of our project complete and ready for testing.

Creating a project

For this project, we'll create a new blank project in Studio. To start a project, complete the following steps:

1. Navigate to the **Start** tab of UiPath Studio.
2. Choose **Process**.
3. Name your project `IT_Helpdesk_Chatbot`, and provide a folder location to save the project to.
4. Click **Create** to create a project and to get started.

That concludes creating a project. We'll revert to the **Main** workflow in a later section where we build out the workflow, but for now, we'll transition into building supplemental workflows.

Building Reset_Password and Upgrade_App workflows

With an automation project created, we can focus on creating some of the supplemental workflows that'll become building blocks of our **Main** workflow. Because we don't have actual accounts to reset or an application license to upgrade, our objective is to create modular workflow sequences that can simulate password resets or application license upgrades.

To start, let's create a `Reset_Password` workflow, as follows:

1. In the `IT_Helpdesk_Chatbot` project, create a new workflow file called `Reset_Password`.
2. Once the new workflow is created, create two arguments, as shown in the following screenshot:

Name	Direction	Argument type	Default value
in_Email	In	String	*Enter a VB expression*
out_SRNumber	Out	String	*Default value not supported*
Create Argument			

| Variables | Arguments | Imports | | 🖐 🔍 100% ▼ |

Figure 10.15 – Reset_Password arguments

The purpose of each argument is this:

- `in_Email`: To input an email address from the chatbot

- `out_SRNumber`: A random SRN that we will output in the chatbot

Once the arguments are added to the workflow, we can start adding activities to the blank canvas.

3. Add a **Message Box** activity, with `"Resetting Password for"+in_Email` as the message and `Info` as the log level. This will output a message box, simulating the steps that automation could take if we were connected to a system that could reset passwords.

4. After the **Message Box** activity, add an **Assign** activity with the following parameters:

- **To**: `out_SRNumber`

- **From**: `"SR"+Right(Now.Millisecond.ToString,2).ToString`

With the **Assign** activity added and using **Right** to generate an almost random number, we have a workflow that looks like this:

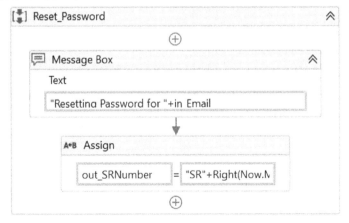

Figure 10.16 – Reset_Password workflow

That concludes the Reset_Password activity. Let's now focus on the Upgrade_App activity.

5. In the IT_Helpdesk_Chatbot project, create a new workflow file called Upgrade_App.

6. Once this new workflow is created, create two arguments, as shown in the following screenshot:

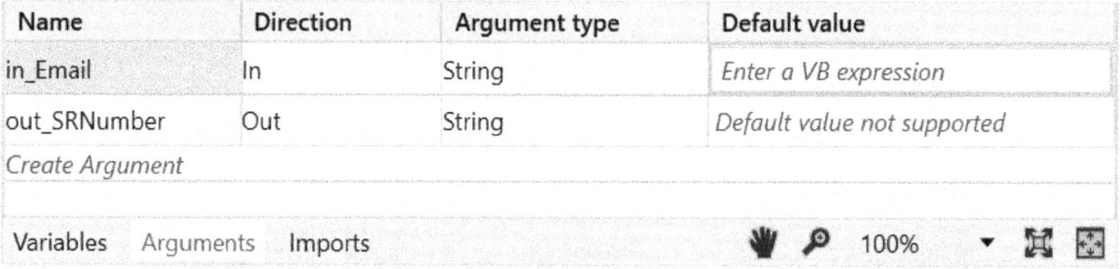

Name	Direction	Argument type	Default value
in_Email	In	String	*Enter a VB expression*
out_SRNumber	Out	String	*Default value not supported*
Create Argument			

Variables Arguments Imports	✋ 🔍 100% ▾ ⛶ 🔲

Figure 10.17 – Reset_Password arguments

The purpose of each argument is this:

- in_Email: To input an email address from the chatbot

- out_SRNumber: A random SRN we will output in the chatbot

Once the arguments are added to the workflow, we can start adding activities to the blank canvas.

7. Add a **Message Box** activity, with "Upgrading App for"+in_Email as the message and Info as the log level. This will output a message box, simulating steps that automation could take if we were connected to a system that could reset passwords.

8. After the **Message Box** activity, add an **Assign** activity with the following parameters:

- **To**: out_SRNumber

- **From**: "SR"+Right(Now.Millisecond.ToString,2).ToString

With the **Assign** activity added and using **Right** to generate an almost random number, we have a workflow that looks like this:

Figure 10.18 – Upgrade_App workflow

That concludes our development of the `Reset_Password` and `Upgrade_App` workflows. In this section, we created workflows that accept a string email body as an input, present an email in a message box activity, generate a random SRN, and output the result. In the next section, we'll work on piecing all this together in our **Main** workflow.

Building the Main workflow

Every automation project starts with a **Main** workflow. The purpose of the **Main** workflow is to kick off the project, invoking other workflows in the process. In this **Main** workflow, we are going to use a `switch` statement to switch between tasks passed by our chatbot. Proceed as follows:

1. Open `Main.xaml` and create three arguments, as shown in the following screenshot:

Name	Direction	Argument type	Default value
in_Task	In	String	*Enter a VB expression*
in_Email	In	String	*Enter a VB expression*
out_SRNumber	Out	String	*Default value not supported*
Create Argument			

| Variables | Arguments | Imports | | 100% | ▼ | | |

Figure 10.19 – Main arguments

The purpose of each argument is this:

- `In_Task`: To input the type of task from the chatbot

- `in_Email`: To input an email address from the chatbot

- `out_SRNumber`: A random SRN we will output in the chatbot

2. Add a **Log Message** activity, with `in_Task` as the message and `Info` as the log level.

3. Add a **Switch** activity, with `in_Task` as the expression.

4. Create a case called `Upgrade AppY`, with an `Invoke` workflow calling the `Upgrade_App` workflow we just created.

5. Create another case called `Password Reset`, with an `Invoke` workflow calling the `Password_Reset` workflow we just created.

The process is illustrated in the following screenshot:

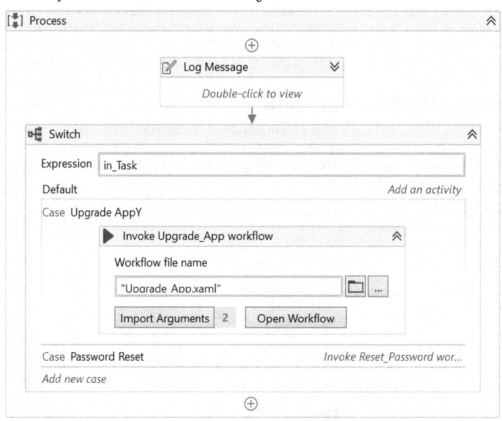

Figure 10.20 – Main.xaml workflow

That concludes configuring the automation workflows of our use case project. In this section, we created our UiPath project and built our supplemental workflows. We then connected our supplemental workflows together within the `Main.xaml` file of our project. In the next section, we will test what we developed.

Testing to ensure stability and improve accuracy

With our development of the chatbot and the automation complete, we can move on to testing both workflows together. Testing any automated workflow before deployment is crucial to ensure that the automation works as expected. In this section, we will publish both the UiPath automation and the Druid chatbot so that we can test the functionality between the two.

Publishing the Druid chatbot

In order for us to be able to test our Druid chatbot and to make it available for users to chat with, we need to publish the bot. Publishing the bot automatically provisions all the required resources in the DRUID environment. A bot is not functional until it is successfully published.

The steps we need to take to publish our chatbot are listed here:

1. Click on the **Settings** icon to the right of **IT Help Desk Bot**, as illustrated in the following screenshot:

Figure 10.21 – Clicking on the Settings icon

2. Once in the **Edit** menu, navigate to the **Details** tab.
3. Click **Publish**.

Once **Publish** is clicked, you should see a notification confirming the request. It may take a few minutes for publishing to complete, but once it does, you can click on the green chat button on the page to test out the chatbot as an anonymous user. Your chatbot should function similarly to this:

Figure 10.22 – The chatbot in action

Feel free to test it out. In the next section, we'll publish our automation so that the Druid chatbot can trigger it.

Publishing the UiPath automation

In previous chapters, we have usually left publishing the automation as a step to be completed after testing. However, in this case, we need to have our automation published so that the Druid chatbot can trigger an automation to run.

> **Important Note**
>
> More information on deploying automation can be found at the following link: https://docs.uipath.com/studio/docs/about-publishing-automation-projects.

For this use case, we're going to quickly publish the automation in our own personal workspace within **UiPath Assistant**, as follows:

1. Within the IT_Helpdesk_Chatbot project, click the **Publish** button in the **Design** ribbon of UiPath Studio.

2. Check that the **Publish to** field is set to Orchestrator Personal Workspace Feed, as illustrated in the following screenshot:

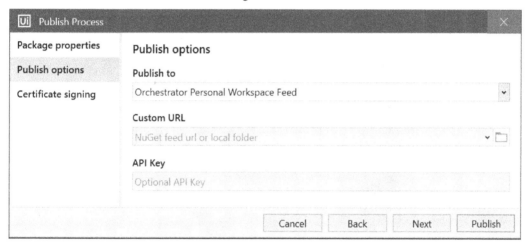

Figure 10.23 – Publishing the automation

3. Click **Publish**.

Once published, you should be able to see the automation within the **UiPath Assistant**. Clicking the **Play** button to the left of IT_Helpdesk_Chatbot will install the automation on your machine and start execution, Let's move to the next section where we will test the chatbot and automation together!

Testing the use case

With our chatbot and automation both published, we can finally start testing our automation! To test out the automation, go back into Druid and converse with your chatbot. Once you click on one of the options to either reset a password or upgrade an application license, you will notice Druid sending a notification that the automation was launched and that the request was successful, as illustrated in the following screenshot:

Figure 10.24 – Launching UiPath from Druid

Because we built an attended automation, Druid communicates with UiPath using UiPath's RobotJS connector. UiPath's RobotJS allows developers to contextually embed local attended robots into applications such as chatbots. When you first run an automation from Druid, you may be prompted by your browser asking whether you want to open **UiPath Assistant**, and once accepted, you may notice a request pop up from **UiPath Assistant**, as illustrated in the following screenshot:

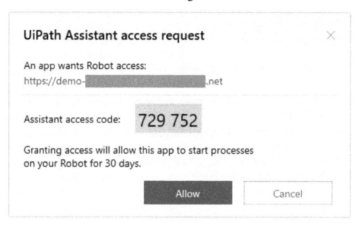

Figure 10.25 – RobotJS prompt

Once you click **Allow**, the automation will kick off, and you should see an SRN returned to the chatbot. Continue to keep testing with a combination of requests. Great job!

> **Important Note**
> If you would like to view a video of the use case in action, navigate to the following link: `https://bit.ly/3DKdXul`

Considerations for production

As we deployed the automation and the chatbot in the last section, we are all set to deploy the automation in our production environment. However, even though our use case is complete, there are some considerations to take into account if you plan to build another use case by leveraging a chatbot.

The first consideration is whether the automation component can be an Unattended Bot. In this use case, we leverage an Attended Robot, but most chatbots should leverage an Unattended Robot when executing the UiPath automation.

By leveraging the **Druid Integration Framework** available on the UiPath Marketplace (`https://marketplace.uipath.com/listings/druid-integration-framework`), you can have Druid trigger an automation by adding transaction items to a UiPath queue item, as illustrated in the following screenshot:

Figure 10.26 – Druid Integration Framework

Adding a queue item will trigger the UiPath unattended automation to begin, completing the process and sending back any relevant information to the Druid chatbot. For future projects, consider leveraging the **Druid Integration Framework** with unattended automation to quickly ramp up an Unattended Automation project.

Summary

In this chapter, we covered building a chatbot use case with *Druid*. In the construction of the use case, we learned how to create a project in Druid, import a template chatbot, build a chatbot flow, and—finally—deploy the chatbot. We also learned how to leverage the connection of the chatbot with UiPath. With these foundations, you can now venture off and build your own cognitive automation.

In the next chapter, we will look into some advanced topics with *UiPath AI Center*.

11
AI Center Advanced Topics

In the previous chapters, we saw how we could leverage **machine learning (ML)** to assist with receipt processing and email classification. While these use cases are very popular, there are many other types of use cases that **Named Entity Recognition (NER)** models or custom ML models can be applied to.

In this chapter, we will cover the following topics:

- **Named Entity Recognition (NER)** with AI Center
- Deploying your own custom models to AI Center

We will introduce the concept of NER, a task that opens new types of use cases we can apply ML to. We will also venture into deploying custom Python ML models into AI Center, as there are many instances where organizations have their own prebuilt ML models that can be leveraged with RPA.

Once complete, we will have an understanding of how to leverage UiPath's NER capabilities, as well as how to connect your own custom Python models to AI Center.

Technical requirements

All code examples for this chapter can be found on GitHub at `https://github.com/PacktPublishing/Democratizing-Artificial-Intelligence-with-UiPath/tree/main/Chapter11`.

Working with UiPath is very easy, and with the Community version, we can get started for free. However, for UiPath AI Center, we will need an enterprise license. You can acquire a 60-day enterprise trial license from `uipath.com`.

With the 60-day enterprise trial, we will have access to the following:

- 5 RPA Developer Pro licenses – Named User licenses include access to Studio, StudioX, Attended Robot, Apps, Action Center, and Task Capture

- 5 Unattended Robots, 5 Testing Robots, 2 AI Robots

- AI Center, AI Computer Vision, Automation Hub, Data Service, Document Understanding, and Insights

For this chapter, you will require the following:

- UiPath Enterprise Cloud (with AI Center)

- UiPath Studio 2021.4+

- `UiPath.MLServices.Activities` package v.1.20.0 or higher

Important Note

Directions on how to install UiPath packages can be found at `https://docs.uipath.com/studio/docs/managing-activities-packages`.

Enabling AI Center in UiPath Enterprise trial

Additional technical requirements include leveraging AI Center. UiPath Document Understanding comes out of the box with the UiPath Community and Enterprise versions, however, with UiPath AI Center, we need to enable the service within the Enterprise trial. You can enable the service by following the given steps:

1. Navigate to **Automation Cloud**: `https://cloud.uipath.com`.

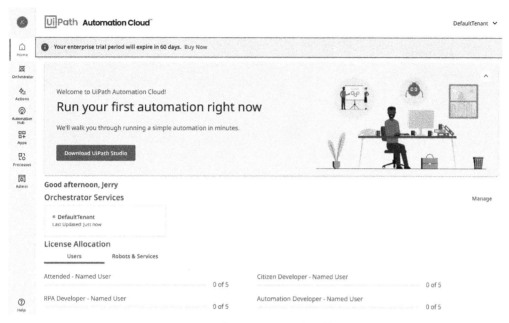

Figure 11.1 – UiPath Automation Cloud home page

2.   Navigate to **Admin**.

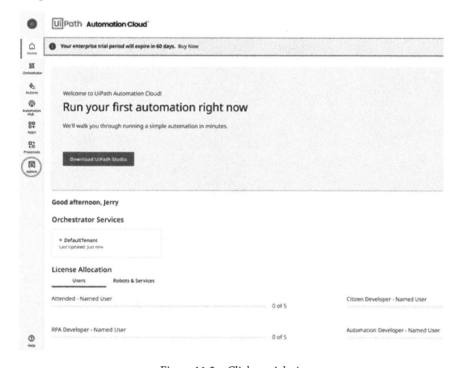

Figure 11.2 – Click on Admin

3.   In **DefaultTenant**, click on the three dots and then click on **Tenant Settings**.

Figure 11.3 – Click on Tenant Settings

4.   Choose **AI Center** under **Provision Services**, and click on **Save**.

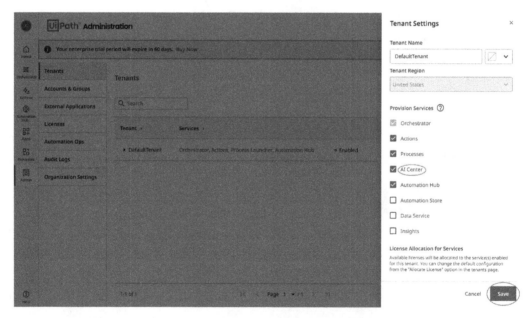

Figure 11.4 – Check AI Center, then click Save

Once AI Center is enabled, we have completed the technical requirements for this chapter. In the next section, we will start gathering context about the use case by reviewing the current state.

NER with AI Center

Many use cases that leverage ML models use some sort of classification. NER is the task of identifying and categorizing key information within texts. Example use cases include extracting and classifying text from emails, letters, web pages, or even transcripts. In this section, we'll discuss AI Center's NER models and discuss how to use the trainable NER model.

Introducing AI Center's NER models

NER is a form of **natural language processing** (**NLP**) that is tasked to process and analyze natural text by detecting and categorizing entities within language. Examples of common entity categories include the following:

- Person: Elvis Presley, Joe Biden

- Company: Google, Apple

- Time: 2010, Winter, 3 a.m.

- Location: New York City, Machu Picchu

Having technology that can recognize these common entities can be extremely powerful when used with automation. UiPath's NER ML model can be used in many ways, as shown in *Figure 11.5*.

Figure 11.5 – Example use cases for NER

These types of use cases cover a lot of automation opportunities, including the email classification use case from an earlier chapter.

To build these use cases, UiPath's AI Center comes with both open source NER models and a UiPath-prepared NER model (*Figure 11.6*).

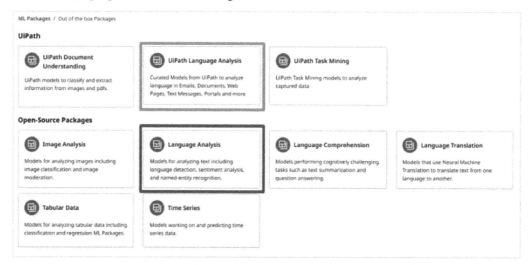

Figure 11.6 – AI Center NER models

Within **Open-Source Packages** > **Language Analysis,** you will find an ML package named NamedEntityRecognition. This ML package is pre-trained to recognize 18 types of different named entities, listed in *Table 11.1*.

PERSON	People, including fictional
NORP	Nationalities, religious, political groups
FAC	Buildings, airports, highways, bridges
ORG	Companies, agencies, institutions
GPE	Countries, cities, states
LOC	Non-GPE locations, mountain ranges, bodies of water
PRODUCT	Objects, vehicles, foods
EVENT	Named hurricanes, battles, wars, sports events
WORK_OF_ART	Titles of books, songs
LAW	Named documents made into laws
LANGUAGE	Any named language
DATE	Absolute or relative dates or periods
QUANTITY	Measurements, as of weight or distance
ORDINAL	"first," "second"
CARDINAL	Numerals that do not fall under another type

Table 11.1 – Entities of the NamedEntityRecognition model

As shown, the `NamedEntityRecognition` open source model can be used to identify many different entities. The one caveat of this model is that it comes pretrained and is not re-trainable. In the next section, we will look into UiPath's custom NER ML model for use cases that require a trainable NER model.

Custom NER with UiPath

For cases when we require an NER model that does not fit into the 18 entities of the open source NER model, or for cases that require retraining, we can use UiPath's out-of-the-box `CustomNamedEntityRecognition` ML model (*Figure 11.7*).

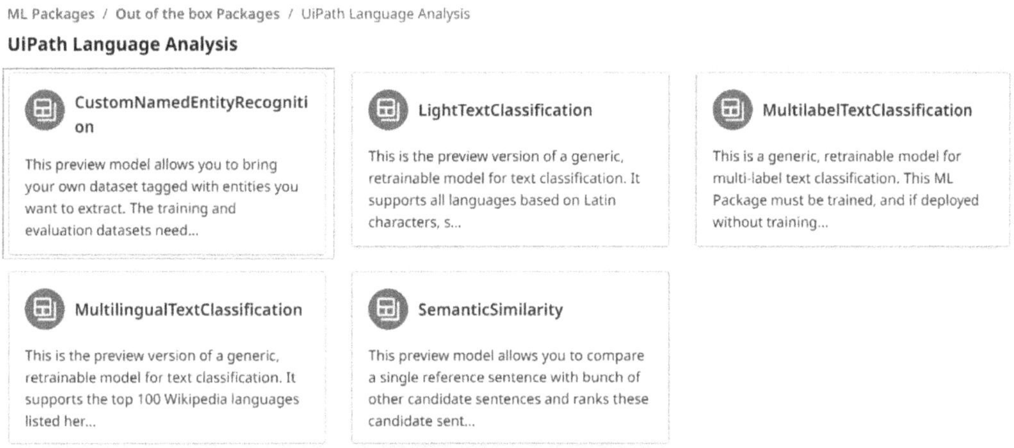

ML Packages / Out of the box Packages / UiPath Language Analysis

UiPath Language Analysis

CustomNamedEntityRecogniti on

This preview model allows you to bring your own dataset tagged with entities you want to extract. The training and evaluation datasets need...

LightTextClassification

This is the preview version of a generic, retrainable model for text classification. It supports all languages based on Latin characters, s...

MultilabelTextClassification

This is a generic, retrainable model for multi-label text classification. This ML Package must be trained, and if deployed without training...

MultilingualTextClassification

This is the preview version of a generic, retrainable model for text classification. It supports the top 100 Wikipedia languages listed her...

SemanticSimilarity

This preview model allows you to compare a single reference sentence with bunch of other candidate sentences and ranks these candidate sent...

Figure 11.7 – UiPath's CustomNamedEntityRecognition out-of-the-box ML model

This ML package allows us to bring our own datasets, pre-tagged (or labeled) with entities we need to extract. This can open the door to many new types of use cases with many different types of entities, such as the following:

- Insurance use case – vehicle entities

- Healthcare – drug entities

- Retail – clothing entities

With this ML model, we need to provide a dataset prelabeled with the entities we want to extract. The training and evaluation datasets need to be in the CoNLL format, which can be performed using a tool named **Label Studio**.

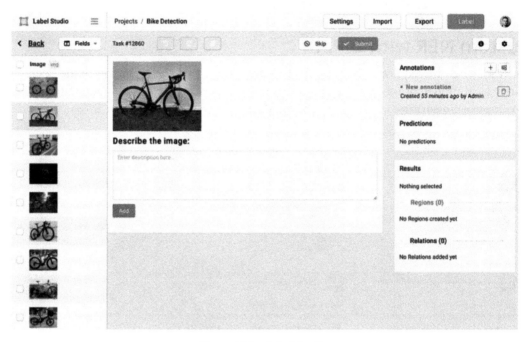

Figure 11.8 – Label Studio

Label Studio is an open source tool that allows you to easily label data types such as audio, text, images, and more. With the `CustomNamedEntityRecognition` ML model, *Label Studio* will be used to export labeled data into CoNLL 2003 format.

> **Important Note**
> For a walk-through on how to use UiPath's Custom NER model, refer to `https://docs.uipath.com/ai-fabric/v0/docs/extracting-chemicals-from-research-paper-by-category`.

Once the datasets are uploaded into AI Center, the creation of an ML skill is the same as we have performed in previous chapters.

In this section, we introduced the concept of NER, learned how it can be relevant to multiple applicable use case scenarios, and learned about the NER models available in UiPath AI Center. In the next section, we will look into deploying custom Python ML models into AI Center.

Deploying your own custom models to AI Center

Our examples in previous chapters have all leveraged out-of-the-box ML models developed and trained by UiPath or the open source community. While these out-of-the-box models may fit most use cases, there may be times when you have a custom ML model that you would like to leverage with automation. UiPath AI Center makes it very easy to deploy pre-made Python or AutoML models. In this section, we'll cover the following:

- Preparing a Python model for AI Center

- Deploying to AI Center

- Interacting with UiPath automation

After covering these points, you should have the skills necessary to deploy your own pre-built models into UiPath AI Center.

Preparing a Python model for AI Center

Before we can deploy a Python model into AI Center, there are a few requirements that we need to add to the Python model. These requirements can be separated into requirements for serving the model and requirements for training the model.

For serving the model, every ML package must contain the following:

- A zipped folder containing a file, `main.py`.

- Within `main.py`, two functions must be implemented:

  - `_init_(self)`: Taking no arguments and loading the model

  - `predict(self, input)`: The function to be called at model serving time, returning a string

- A file named `requirements.txt` with the dependencies needed to run the model.

If we wanted to add re-training and evaluation components to the ML model, the following requirements are necessary:

- A file named `train.py` within the same directory as `main.py`.

- Within `train.py`, the following functions must be implemented:

- `_init_(self)`: Taking no arguments and loading the model

- `train(self, training_directory)`: The function to be called when training a training pipeline run

- `evaluate(self, evaluation_directory)`: The function to be called during an evaluation pipeline run

- `save(self)`: The function to be called after training time to persist the model

- `process_data(self, input_directory)`: The function to be called during a full pipeline run

> **Important Note**
>
> Additional information on preparing Python code for AI Center ingestion can be found at `https://docs.uipath.com/ai-fabric/v0/docs/building-ml-packages`.

For this example, we're only going to create a simple ML model that is pretrained, hence only requiring `main.py` and not `train.py`. To take a look at the example Python file prepared for AI Center, navigate to `https://github.com/PacktPublishing/Democratizing-Artificial-Intelligence-with-UiPath/blob/main/Chapter11/Linear_Regression_Model/main.py`.

This example Python file, `main.py`, is a simple linear regression model that has been adapted to conform with AI Center's requirements. As per AI Center's requirements, the code contains the following:

- A class called `Main`

- A function, `_init_(self)`, that initializes the model

- Another function, `predic(self, input)`, that runs the linear regression model based on an input, and returns an output string

> **Important Note**
>
> More details on the source code for the Python example can be found at `https://towardsdatascience.com/simple-machine-learning-model-in-python-in-5-lines-of-code-fe03d72e78c6`.

With the `main.py` file conforming to the specifications required by AI Center, one additional requirement is necessary before we can deploy to AI Center. We must create a text file, `requirements.txt`, listing the Python packages that are required for the project. For our example Python file, the `requirements.txt` file would look like this:

```
numpy==1.22.2
scikit-learn==1.0.2
scipy==1.7.3
joblib==1.1.0
threadpoolctl==3.1.0
```

This file lists the five Python dependencies necessary in order to run our project. Both these files (`main.py` and `requirements.txt`) should be placed into the same folder.

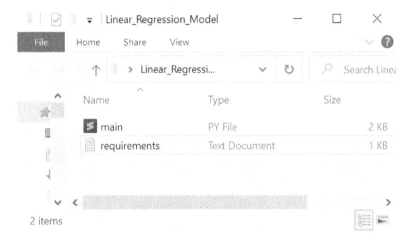

Figure 11.9 – The Linear_Regression_Model folder

In this example, both files have been placed into a folder named `Linear_Regression_Model`. This folder should then be compressed to be ready for AI Center. In the next section, we will upload the ZIP file to AI Center and create the ML skill.

Deploying to AI Center

Now that our Python model is created and our project folder is zipped, let's continue by deploying the project to AI Center. The steps to deploy the package to AI Center are very easy, and are as follows:

1. Open AI Center and create a new project or leverage an old AI Center project.
2. Navigate to **ML Packages** in AI Center.

3. Select **Upload zip file** to upload our prepared zip file.

4. When presented with the screen to create a new package, provide a package name, description, input, the Python language, and choose the zipped package.

Our Python package should appear similar to *Figure 11.10*:

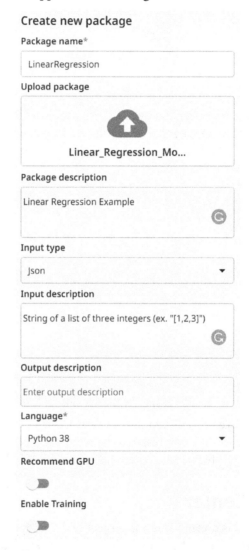

Figure 11.10 – Uploading a package to AI Center

When uploading the package to AI Center, be sure to choose the correct version of Python used when developing the model. Because we did not include a training component, **Enable Training** is disabled.

Once the package is uploaded, it will have an *undeployed* status. Our next step is to deploy the package by creating an ML skill:

1. Navigate to the **ML Skills** tab of AI Center.

2. Click **Create New** to create a new ML Skill.

3. When presented with the **Create New ML Skill** screen, provide the following:

    - A skill name

    - The `LinearRegression` package

    - The most recent package version

Our ML Skill should appear similar to *Figure 11.11*:

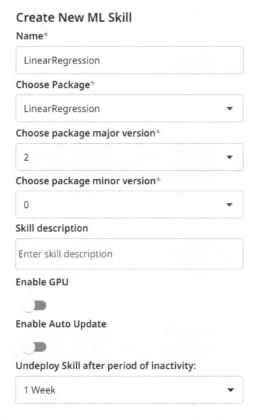

Figure 11.11 – Creating a new ML Skill

Once deployed, the ML Skill should show a status of *Available*, signaling a successful deployment. In the next section, we will create a new UiPath project and connect it to our newly created ML skill.

Interacting with UiPath automation

With the ML Skill deployed into AI Center, the last step of the example is integrating it with UiPath automation:

1. Open UiPath Studio and create a new blank project.

2. Add the `UiPath.MLServices.Activities` package through the **Manage Packages** button.

3. Add an *ML Skill* activity into the workflow, with the following parameters:

    *   **Connection Mode**: Robot

    *   **Skill**: LinearRegression

With the connection to the ML Skill created, we can now test it out by passing inputs, as shown in *Figure 11.12*.

Figure 11.12 – Testing the ML Skill

And by providing the input [10,20,30], we can see a result of 140 and three coefficients.

And that's it! In this section, we prepared a custom Python ML model for AI Center, zipped and uploaded the package to AI Center, and created an ML Skill and connected it to UiPath. With what we covered, you should be able to prepare and import a custom ML model into AI Center.

Summary

In this chapter, we introduced the concept of NER and how it can be applied with UiPath AI Center. We then continued on to learn about how to prepare custom Python models for AI Center by walking through a simple use case with a linear regression Python model. With these foundations, you can now venture off and build use cases with NER or upload your own custom Python models into AI Center.

And with that, we have reached the conclusion of the book. At this point, we believe that you have the skills and foundation to identify, design, build, and test cognitive automation opportunities. We hope you enjoyed the journey and can take your learnings to new heights as we bring cognitive automation to the world.

Index

Packt.com

Subscribe to our online digital library for full access to over 7,000 books and videos, as well as industry leading tools to help you plan your personal development and advance your career. For more information, please visit our website.

Why subscribe?

- Spend less time learning and more time coding with practical eBooks and Videos from over 4,000 industry professionals

- Improve your learning with Skill Plans built especially for you

- Get a free eBook or video every month

- Fully searchable for easy access to vital information

- Copy and paste, print, and bookmark content

Did you know that Packt offers eBook versions of every book published, with PDF and ePub files available? You can upgrade to the eBook version at packt.com and as a print book customer, you are entitled to a discount on the eBook copy. Get in touch with us at customercare@packtpub.com for more details.

At www.packt.com, you can also read a collection of free technical articles, sign up for a range of free newsletters, and receive exclusive discounts and offers on Packt books and eBooks.

Other Books You May Enjoy

If you enjoyed this book, you may be interested in these other books by Packt:

Learn Amazon SageMaker - Second Edition

Julien Simon

ISBN: 9781801817950

- Become well-versed with data annotation and preparation techniques
- Use AutoML features to build and train machine learning models with AutoPilot
- Create models using built-in algorithms and frameworks and your own code
- Train computer vision and natural language processing (NLP) models using real-world examples
- Cover training techniques for scaling, model optimization, model debugging, and cost optimization
- Automate deployment tasks in a variety of configurations using SDK and several automation tools

Machine Learning Engineering with Python

Andrew P. McMahon

ISBN: 9781801079259

- Find out what an effective ML engineering process looks like
- Uncover options for automating training and deployment and learn how to use them
- Discover how to build your own wrapper libraries for encapsulating your data science and machine learning logic and solutions
- Understand what aspects of software engineering you can bring to machine learning
- Gain insights into adapting software engineering for machine learning using appropriate cloud technologies
- Perform hyperparameter tuning in a relatively automated way

Packt is searching for authors like you

If you're interested in becoming an author for Packt, please visit `authors.packtpub.com` and apply today. We have worked with thousands of developers and tech professionals, just like you, to help them share their insight with the global tech community. You can make a general application, apply for a specific hot topic that we are recruiting an author for, or submit your own idea.

Share Your Thoughts

Now you've finished *Democratizing Artificial Intelligence with UiPath*, we'd love to hear your thoughts! Scan the QR code below to go straight to the Amazon review page for this book and share your feedback or leave a review on the site that you purchased it from.

https://packt.link/r/1-801-81765-0

Your review is important to us and the tech community and will help us make sure we're delivering excellent quality content.

www.ingramcontent.com/pod-product-compliance
Lightning Source LLC
Chambersburg PA
CBHW062048050326
40690CB00016B/3023